TRANSMISSION
OF RADIANT ENERGY

BY

OPHTHALMIC GLASSES

Being an Essay Contributed

to

The American Encyclopedia of Ophthalmology

by

CHARLES SHEARD, A.B., A.M., Ph.D.

Physiological Opticist, The American Optical Company; formerly Professor and
Director and now Non-resident Professor of Applied Optics, The Ohio State
University; Collaborator in *The American Encyclopedia of Ophthalmology;*
Member of the Optical Society of America, The Physical Society,
Honorary Member of the American Optometric Association, etc.;
Author of *Physiological Optics* (1918), *Dynamic Ocular Tests*
(1917), *Dynamic Skiametry* (1920), *Ocular Accommodation*
(1920) and series of articles on *Mathematical Studies in
Optics* (1910-11), *Cylindrical Lenses* (1914), and
several contributions on The Ionization from
Hot Salts and Metals, The Free Vibra-
tions of Lecher Systems, etc., Edi-
tor of *The American Journal of
Physiological Optics*

Illustrated with Seventy-nine Diagrams

Chicago
CLEVELAND PRESS
1921

DEDICATED TO
ALL THOSE SCIENTIFICALLY
INTERESTED IN THE PRACTICES
OF OCULAR REFRACTION

TRANSMISSION OF RADIANT ENERGY

BY

OPHTHALMIC GLASSES

TABLE OF CONTENTS

483432

FOREWORD

———

Through the courtesies of Dr. Casey A. Wood, Editor-in-chief of the *American Encyclopedia of Ophthalmology*, and the publisher, Dr. Geo. Henry Cleveland, of the Cleveland Press, the writer of this brochure has been able to secure a reprint of the original essay for the use of students and practitioners who may be interested in having at first hand data relative to the transmission of radiant energy by ophthalmic glasses.

A considerable number of pages have been devoted to a discussion of the nature and distribution of radiant energy and various common methods of producing and investigating ultraviolet, visible and infrared radiations, in order that the full significance of the data presented in the remainder of the essay may be disclosed.

The writer claims for these pages nothing more than an attempt to put together and present some of the salient points with reference to a subject which is of considerable interest and importance. Much has yet to be discovered in the field of the relations between radiant energy and the eye.

CHARLES SHEARD

Research Division,
American Optical Company,
Southbridge, Mass., 1921.

TRANSMISSION OF RADIANT ENERGY

BY

OPHTHALMIC GLASSES

CHAPTER I. RADIANT ENERGY.

We know that all about us energy is constantly being transferred from one place to another. When this occurs something is moving and something is being moved through, hence there must be a medium of transmission. Radiation is a form or kind of energy and therefore can be produced from other forms of energy and can be converted into other kinds. Heat energy is a convenient form for the production of radiation. For instance, the resistance offered by the filament of an incandescent lamp to the flow of electric current produces energy which is radiated out as disturbances in the ether. These disturbances, having varying wavelengths, are characterized as heat and infra-red, visible, or ultra-violet light when they are intercepted and received by a suitable body. Heat, for example, is not radiant energy but exists only when it ceases to be radiation. And again, radiations of the proper wavelengths, when received by the eye, cease to exist as radiant energy and become what is popularly spoken of as light.

Nature and Distribution of Radiant Energy.

Light is physically defined as a periodic or rhythmic electromagnetic disturbance in a transmitting medium, the ether, traveling in the form of transverse waves with a velocity of approximately 186,000 miles per second. At first glance it is not evident that there is any connection between light and electricity (or electric waves). Such a relation was predicted mathematically by Maxwell at about 1870. In this theory the assumption was made that light waves are identical with electromagnetic disturbances which are given out from a body in which electrical oscillations are occurring. Hertz later produced these electric waves and the theory of Maxwell was given an experimental verification and the science of wireless telegraphy and telephony was born. The oscillating molecules, atoms and their electrons are presumably responsible for these pulses of electromagnetic energy. While all these radiations travel in free space with the same velocity, they differ considerably in their velocities in ordinary media: in glass, for example, the violet rays travel less rapidly than the red rays. All of these rays carry energy; that is, the rays are the actual physical energy in transmission and hence when absorbed will produce heat, chemical action or physiological change. Sometimes the absorbed radiation is not converted wholly into heat but enters into chemical reaction or is changed into radiant energy of wavelength differing from that of the absorbed energy. Rays shorter than the visible and known as ultra-violet are very active chemically, affect photographic films, destroy bacteria and animal tissues such as the outer membrane (the conjunctiva) of the eye: they also cause phosphorescence and fluorescence. The long wavelengths of the visible spectrum are received by the normal retina and interpreted as representing light and redness of color. Beyond the longest visible red ray (in the region of 7600 Angstroms or tenth-meters) come the infrared rays. These are commonly spoken of as the heat rays because of the fact that the energy of the radiations is transformed ordinarily in the maximum percentages into heat.

Wavelengths of Light.

The wavelength of light can be measured with extremely high accuracy (*vide* the experiments of Michelson with the interferometer and so forth) and has been proposed as the absolute standard of length instead of the meter, which was intended to be a ten-millionth part of the earth's quadrant. It is found, however, that different kinds of radiant energy have widely different wavelengths; for example, the different colors (as we may call them) of light, have different wavelengths, red light having the longest and blue the shortest length of the visible region. The wavelength may be specified in units known as the micron (μ), the tenth-meter (t. m.) and the Angstrom (A). The micron is the one-millionth part of a meter or the one-thousandth part of a millimeter. The Angstrom unit and the tenth-meter are each equal to one ten-millionth of a millimeter. As a result a radiation, which may be specified as red light, may be defined as 7000 A., 7000 t. m. or 0.7 μ.

Light is, therefore, a form of radiant energy which, when received by the eye, gives normally the sensations of sight and of color. The visible spectrum spans one octave practically. The question as to the sources of these disturbances in the ether leads us to a brief statement of the fact that atoms are now known to be composed of infinitely small particles which are called electrons. These electrons, under given physical conditions, oscillate or vibrate with certain definite periods or frequencies. The number of electrons composing the atom and the rate of frequency of vibration depend upon the element, or the so called atomic number. In order to make that about which we are writing the more tangible, imagine a carbon arc set-up and ready for operation. Before the arc is struck there is no light or heat and no changes in the appearance of the carbon. As the arc is "struck," electric current flows through the carbons and they become incandescent, throwing out heat and light and other forms of energy. By virtue of the disturbances set up in the carbon electrodes under the influence of an electrical potential or pressure, the molecules and atoms of the carbon are agitated: the current passing through the carbon heats it and as the atoms and molecules are thrown into vibration and absorb energy the electrons composing the atom become correspondingly disturbed, vibrating with greater frequencies and by their vibrations and collisions with each other produce disturbances in the ether. These disturbances are propagated into space by wave motion and since the disturbances within the carbon are very complex, a corresponding complex emission of wave energy follows. When the frequency of the waves approaches 400 million millions per second (4×10^{14} cycles per second) a dull red glow appears. When the frequencies of the waves range between 400 and 800 million millions per second (4×10^{14} to 8×10^{14} cycles) the sensation produced in the eye is that of white or nearly white dependent upon the percentages of the various wavelengths. The range between 400 and 800 million millions repesents the visual spectrum between violet (800) and the extreme red (roughly 400). The colors of the so called visible spectrum are as follows:

	Frequency.	Wave-length.
Red............	400×10^{12} cycles	75×10^{-6} cm.
Orange	460	65
Yellow	508	59
Green	566	53
Blue	652	48
Indigo	710	45
Violet	800	40

It is to be noted at this point, however, that the actual limits of the so-called visual spectrum to the dark adapted eye are not to be set at 0.75μ and 0.4μ respectively.

Spectrum of Radiation.

The spectrum of radiant energy extends from the shortest wavelength known, that of the X-ray having a wavelength of the order of magnitude of 1×10^{-8} cm. (one hundred-millionth part of a centimeter) to the longest waves due to the fields of alternating current circuits and having wavelengths approximating fifteen thousand miles. Within the past few years methods of crystal spectrometry as devised by Laue and elaborated experimentally by the two Braggs have led to determinations of the wavelengths of X-rays. Shaw has made measurements upon the γ rays given out by radioactive substances and has found them from ten to one hundred times shorter in length than are the hardest Roentgen radiations. Several octaves (an octave representing a doubling of the frequency) are missing between the longest X-ray and the shortest so called ultra-violet wavelength which was set a year or so ago by Lyman, of Harvard, at about 0.06μ or something less than 0.1μ. Just as these words are being penned, however, comes the report that Millikan, of Chicago, has succeeded in getting ultra-violet waves about ten times shorter than the shortest obtained by Lyman. The greater part of the research conducted in this region has been done by means of photography carried on with spectroscopes and specially sensitive plates enclosed in vacuum chambers. Ordinary crown and flint lenses and prisms cannot be used in such experimentation, for they absorb up to wavelength 0.30μ as about the lowest limit. Quartz lenses and prisms are therefore used. In working with the spectrum below 0.185μ such effects as absorption by the quartz and absorption by the gelatine of the photographic films begin to exert their influence.

The actual division line between the ultra-violet and the visible spectra is commonly and somewhat arbitrarily set at 0.4 μ. Doubtless this division point has arisen because of the various determinations of the "visibility curves" of eyes. Hyde, Cady, Forsythe, Hartwell, Nutting, Ives, Reeves, Coblentz and others have investigated these visibility curves with considerable care: something in the neighborhood of 0.4 μ in the violet is about the limit of accurate investigation with present experimental photometric devices. These visibility curves are not to be confused with the shortest or longest wavelengths *per se* which a dark-adapted eye can see and measure. Certain data to be presented later indicate the transmission of radiant energy through the ocular media to the retina having wavelengths considerably shorter than 0.4 μ. There are doubtless crystalline lenses which absorb all wavelengths shorter than 0.4 μ: but Hallauer, for instance, found many who could see as low as 0.36 μ to 0.37 μ and claims that in the case of youthful lenses there is an actual transmission of rays of wavelength 0.31 μ to 0.33 μ. At the time these words are being written a careful and detailed experimental investigation is being conducted by the writer and his colleagues upon the lowest limit of radiation visible to a dark-adapted eye. Wavelength 0.34 μ (with definite and yet very faint violet color) has been readily observed by several and 0.32 μ under proper conditions of intensity. It does not seem logical, therefore, to set the limit of visible radiation as high as 0.4 μ. We shall, however, until the weight of evidence is to the contrary, follow the arbitrary division already made, although in certain curves (*vide* Figs. 27-35) we have set the region at about 0.37 to 0.38 μ.

The visible spectrum covers about an octave and extends, roughly, from 0.38 μ to 0.8 μ. It comprises a very small portion of all the known spectrum. The highest sensitivity of the eye is in the yellow-green region at 0.56 μ practically. The extreme visible red is at about 0.79 μ. Hyde and his collaborators have investigated the "visibility" curves in this region (red end) with considerable thoroughness.

The red end of the spectrum is at the beginning of the ultra-red or infra-red radiations. These rays are often, but improperly, spoken of as heat rays, for they should properly be classed and spoken of in the same manner as are the ultra-violet and the visible. In 1800 Sir William Herschel found that when a thermometer with a blackened bulb was moved into the spectral region just beyond the red, there was a rise in temperature indicated. This proved that there were radiations beyond the limit of visual sensitiveness. Sir William Abney succeeded in photographing the infra-red spectrum out to 1.1 μ with specially prepared plates. The late Professor Langley, of fame in these days for his experimentation upon submarines and aeroplanes, constructed an instrument known as a bolometer. In this instrument, based upon the Wheatstone bridge, one arm consists of a fine blackened platinum wire or grid. When this receives radiation it absorbs it and the temperature is raised, the resistance of the wire changed and a current produced in a detecting instrument, a galvanometer. Langley plotted the spectrum to about 61 μ. By the methods of "reststrahlen," interference and focal isolation various experimenters, including such names as Rubens, Hollnagel, Nichols, Trowbridge, and Wood and his co-workers, have succeeded in extending the investigations step by step into the infra-red region to be-

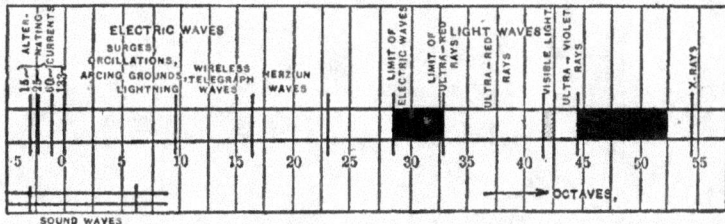

Fig. 1—The spectrum of radiation. (From Steinmetz: *Radiation, Light and Illumination.*)

tween 200 and 300 μ. This corresponds to a wavelength of about 0.2 to 0.3 mm. Rubens and von Baeyer in 1911 found a maximum in the long wave radiation from a quartz mercury arc at 343 μ.

About four octaves gap exists between the longest infra-red radiation as detected by the method of focal isolation and the shortest Hertzian or electric wave thus far found by von Baeyer and having a length of about two millimeters. In passing from one set of radiations to the other we are passing from the region where the *molecule* constitutes the minimum sized vibrator to that in which *molar* relations hold, for electric and Hertzian waves are produced by discharges between electrodes. High frequency currents, surges and oscillations, arcing, wireless, lightning phenomena and so forth, have wavelengths ranging from the limiting wavelength of a fraction of a centimeter as determined by von Baeyer up to 4,000,000,000 μ in length or, in other words, miles in length. Alternating current fields have cycles varying from 15 to 133. Such figures give us as wavelengths something of the order of 13,000 miles and 1400 miles respectively.

Figure 1 is a graphical reproduction from Steinmetz (*Radiation,*

Light and *Illumination*, page 18). It is not, at some points, up to date in its representation, but it serves in a very satisfactory manner to graphically illustrate the distribution of radiant energy from alternating currents to X rays.

Radiation and Light Sensation.

The distribution of the energy among the different wavelengths given out by an incandescent solid is shown in Figure 2. This curve is known as a radiation curve and is the envelop of the plotted values of the energy over a great range of wavelengths. A continuous spectrum, in contradistinction to a line spectrum such as is given by a mercury arc for example, is characteristic of the radiation from solid bodies. In the curve of Figure 2, the height of the curve at any point above the base line is a measure of the relative amount of

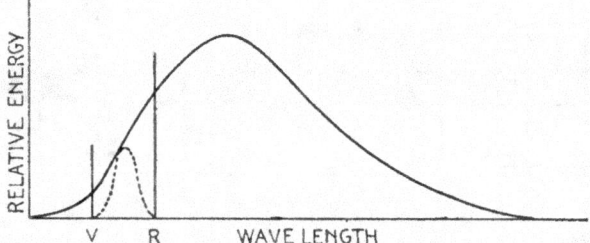

Fig. 2—Radiation curve of an incandescent solid. (Courtesy of M. Luckiesh.)

energy possessed. The amounts of energy for various wavelengths are by no means equal. The region to which the eye responds or is "tuned" lies between V (violet) and R (red). This region, following the diagram taken from Luckiesh (*Color and Its Applications*, page 8), is exaggerated in extent for the sake of clearness.

Figure 3, copied from the work of Langley, shows the relative distribution of energy in the spectra of the gas flame, the electric arc, the solar spectrum and the fire-fly.

As the temperature of an incandescent body is increased, the energy contained in the shorter wavelengths increases more rapidly than the energy in the longer wavelengths. In the visible spectrum the violet and the adjacent rays increase in intensity more rapidly with increase of temperature than does the red end. This causes the light emitted by an incandescent solid to become bluer in color (or less red, since the redness is the noticeable feature) as the temperature is increased. The effect of raising the temperature on the distribution of radiant energy given out by an incandescent solid is shown in the curves given in Figure 4. The numbers on the curves indicate the absolute black-body temperatures (i. e. above —273° C, since 0° C=273 K.). The

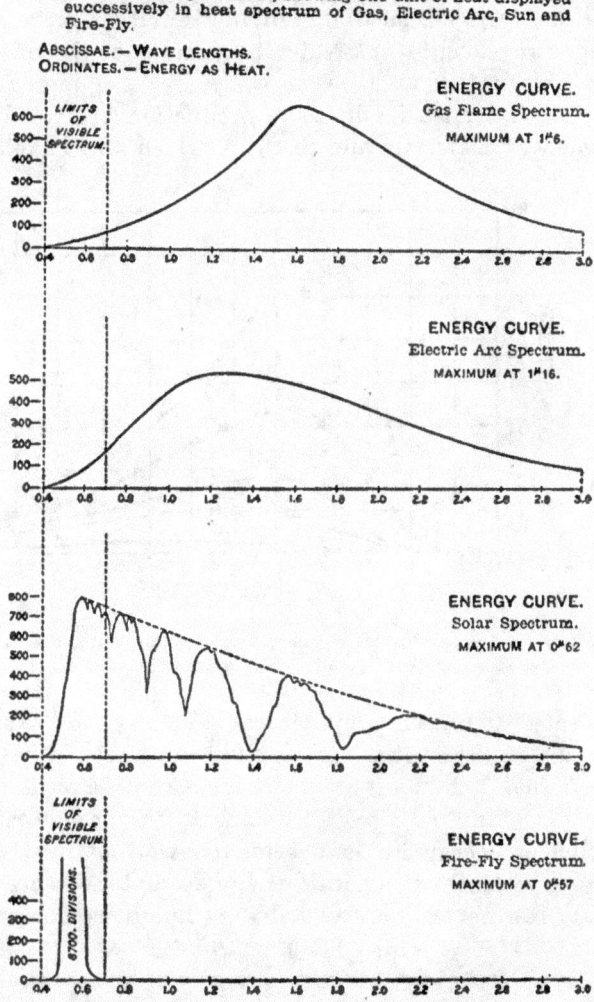

Fig. 3—Distribution of energy in spectra of gas flame, electric arc, sun and fire-fly.

wavelengths are in μ; the rays to which the eyes are sensitive are enclosed between V and R. Thus, as the temperature is raised, the maximum of the radiation curve shifts toward the shorter wavelengths.

The energy distribution curve for sunlight (Figure 3) shows that the maximum lies in the visible region. This has brought forward the hypothesis that the eye, being as we know it to be the product of the processes of evolution, has become most sensitive to the rays of such wavelengths as are in maximum percentages in the radiation from the sun. As the maximum of the radiation curve shifts toward the shorter wavelengths, a greater proportion of the energy is found in the visible region and this accounts for the increased luminous efficiency. All tendencies in light production are to the end of the development of materials and methods which will enable the sources

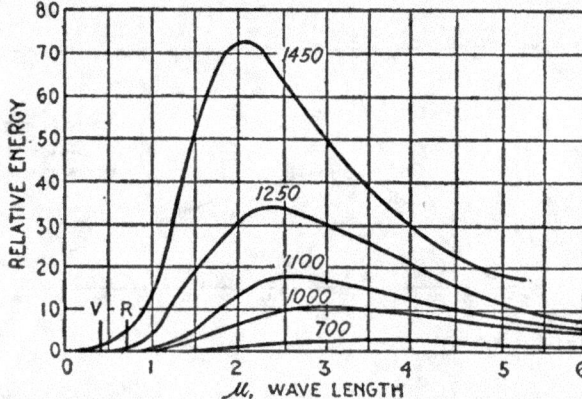

Fig. 4—Showing the effect of temperature on the radiation from an incandescent solid (black body). (Courtesy of M. Luckiesh.)

to be operated at higher temperatures. This is the advantage of the tungsten filament over the carbon filament lamp. The ideal source from the visual standpoint would be one which affords a continuous spectrum extending only from the blue to the orange roughly. The distribution of energy in the spectrum given out by the fire-fly approaches very closely this ideal. Langley and Coblentz have shown that ninety-five per cent. is available as luminous energy.

CHAPTER II. COMMON METHODS OF PRODUCING AND INVESTIGATING ULTRA-VIOLET, VISIBLE AND INFRA-RED RADIATIONS.

Spectrographs.

Spectrographs and spectrometers are the instruments commonly used in investigations upon those regions of the ultra-violet, visible and infra-red which are of interest to us either from the standpoint of

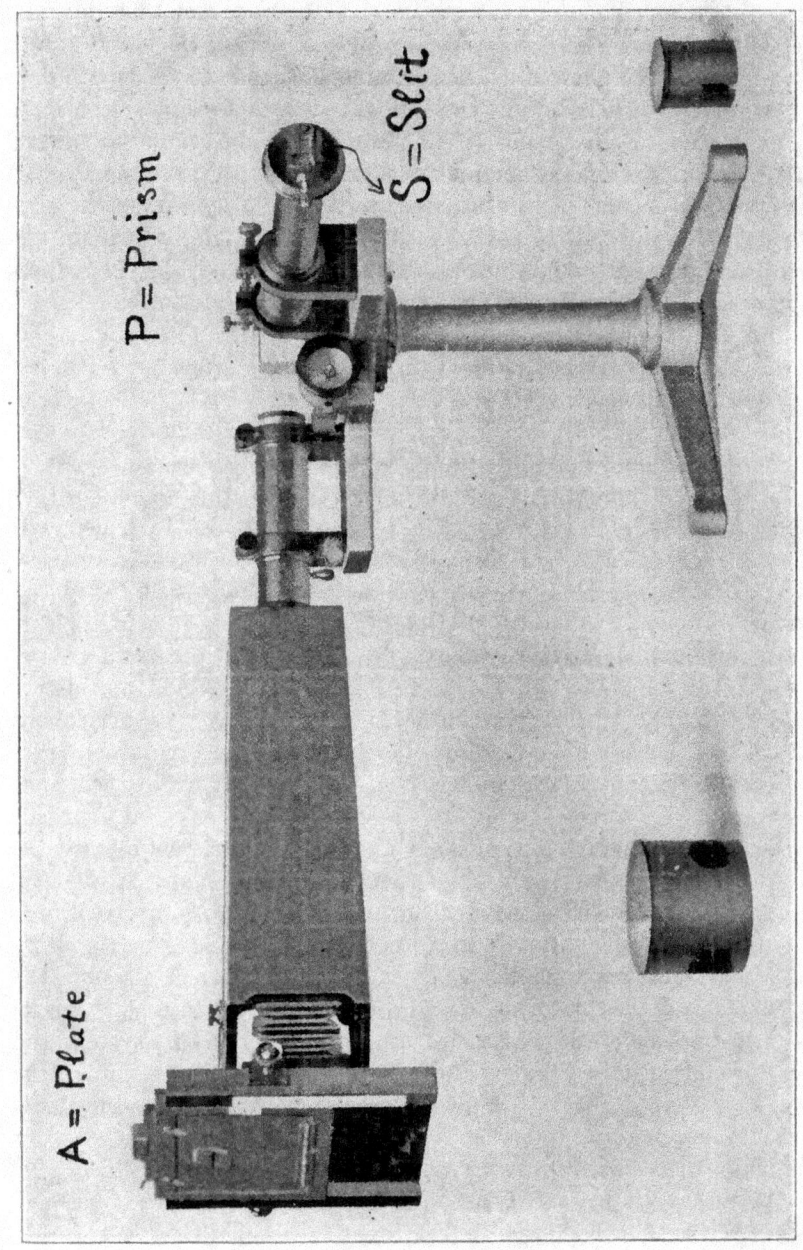

Fig. 5—A form of modern spectrometer.

the eye or of ophthalmic glass transmission. One of the modern forms of spectrograph is shown in Figure 5 and with the use of this diagram the essential features of this *prism* spectrograph will be pointed out. The three essential parts of such instruments are the collimator, the prism system and the photographic apparatus or telescope which may take its place in visual work. The source of light is placed before a narrow slit S in the collimator tube, or the light from the sun or an electric arc may be concentrated by means of a lens upon the slit. At the other end of this tube is placed an achromatic lens system and the slit and lens are so adjusted that parallel light will fall upon the prism P. In passing through the prism the light suffers dispersion and as a result there emerges a parallel beam of red light and a parallel beam of violet light with beams of the other wavelengths situated between them. The pencil of red light is brought to a focus upon the photographic plate A by means of an achromatic lens system in the telescopic tube: the violet rays as well as the remainder of the visible spectrum are likewise focused. Since the dispersion for these rays is different a spectrum, extending from red to violet, will be found upon the plate or visually obtained at the eye-piece of the telescope. In order that wavelengths may be determined the prism must be calibrated. One of the latest makes of spectrometers carries the constant deviation prism and is provided with a drum device, the barrel of which is calibrated in wavelengths. Such an instrument as this, or of a similar character, carrying a glass prism and glass lenses permits of the making of photographic or ocular observations of the visible spectrum. When, however, it is desired that investigations be carried on in the *ultra-violet*, all lenses and prisms must be of *quartz*, since this substance does not absorb the ultra-violet until the limit of about 1800 Angstroms is reached.

Other types of instruments employ the principle of auto-collimation. By this method the collimator and camera lenses are entirely suppressed, the only optical member being the prism itself. The employment of the principle of auto-collimation with a 30° prism simultaneously shortens the apparatus, simplifies the lens system and avoids trouble due to the rotatory properties of quartz, since the prism is traversed twice in opposite directions. The necessary condition for a pure spectrum is that all incident and refracted rays shall make the same angle with the refracting surface. This is accomplished by giving the front and back surfaces of the prism P (Figure 6) suitable spherical curvatures.

An excellent instrument to use for the photographic examination of the ultra-violet end of the spectrum is the Fery quartz spectrograph. This instrument is shown diagrammatically in Figure 6. The quartz prism P is silvered on the back and is ground in such a way that radiation received from a source in front of the slit at A is brought to a focus at E after reflection from the back of the prism. Since the prism in this instrument performs the functions of both the lenses and the prism in an ordinary spectrograph, there are no losses due to absorption other than those which occur in the prism.

The matter of a satisfactory source of illumination for work in the ultra-violet region is worthy of more than a passing remark. Iron arcs and similar devices are fairly rich in ultra-violet but are not continuous. The iron arc spectrum on direct current is redundant in lines; the same arc with condensers used across the arc affords a satisfactory source and a very uniformly and continuously burning

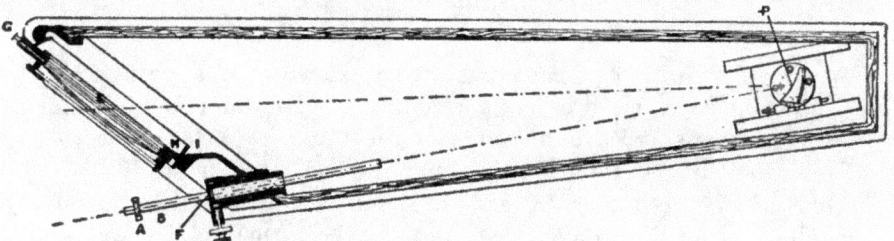

Fig. 6—The Fery quartz spectrometer.

one. A condensed spark discharge, using an induction coil, one electrode being made of iron and the other of an alloy of cadmium, aluminum, magnesium and zinc gives a most excellent spectrum for work in the ultra-violet; while it is fundamentally a line spectrum it has superposed upon it rather extensive portions of a continuous spectrum down to about 2300 t. m.

Another device used by some investigators upon the subject of glasses for protecting the eyes consists of reflecting the light from a mercury arc from the face of a magnesia block. The spectrum is, of course, discontinuous. In its use, however, a spectrum is formed by a small quartz spectrograph, the slit being wide enough to furnish bands from each ultra-violet line of sufficient width for photometerings. A series of exposures of equal length but with different known illuminations of the magnesia surface can be made and a comparison with the transmission through the media under examination carried out. The photographs as thus made can be measured for density on a polarization photometer or other device of a similar character and curves can be plotted showing the connection between the density and illumination for each line. It is thus possible to determine the transmission of glass for each wavelength obtainable from the source of illumination.

Perhaps the best device for producing a continuous spectrum is that consisting of two electrodes—aluminum or brass, for example—under water and actuated by high frequency. A wireless oscillation transformer as a source of excitation has been found very satisfactory. There is less trouble with the question of purity of the water used but it does not afford quite as continuous a spectrum. The limit of continuity is, however, about 2100 t. m.; there is a gradual falling off in intensity of the rays in the extreme ultra-violet. Details in this matter are given by Howe in a paper on a Photometric Method of Measuring Ultra-violet Absorption (*Phys. Review*, Ser. II, Vol. 8, 1916).

Infra-red Spectrum.

Herschel in 1800 demonstrated the existence of a portion of the spectrum beyond the extreme visible red. He did this by means of a thermometer with a blackened bulb. A rise in temperature was indicated in the region just beyond the red. In the study of infra-red rays we need a prism made of a substance which does not absorb radiations of long wavelength and, in the second place, we need a device or receiving instrument which will indicate a very small rise of temperature due to the radiations absorbed. Prisms of rock-salt (or fluor spar: not now used much because of the scarcity) are employed. The instruments used for absorbing the radiations and indicating the consequent rise in temperature are the thermopile, the radio-micrometer and the bolometer.

Dr. Langley of the Smithsonian Institute carried on a very elaborate study of the infra-red solar spectrum making use of his bolometer. This instrument consists of blackened strips of platinum, about a tenth millimeter in breadth and a hundredth millimeter in thickness, arranged to form two arms of a Wheatstone bridge. When the usual galvanometer and battery connections are made, the resistances in the remaining arms are adjusted so that the galvanometer shows no deflection. When radiation falls upon one arm of the bridge, however, the balance of the bridge is destroyed and a deflection of the galvanometer follows. This galvanometer deflection affords a means of relative measurement of the absorption of energy. The sensitiveness of the instrument is such that a rise of temperature amounting to not more than one hundred-millionth of a degree Centigrade can produce a measurable deflection. "What would be a dark band in the spectrum, could our eyes be affected by the long infra-red waves, will fail to heat the platinum strip and the galvanometer deflection will be diminished or reduced to zero" (*Edser.* Light, page 345). By means of this bolometric device Langley investigated the infra-red through a region extending from 0.76 μ (7600 t. m.) to 5.3 μ (53,000 t. m.).

In 1880 Sir William Abney obtained through the use of specially prepared plates a photographic record out to 1.1 μ. The special character of plates and the difficulty of handling photographic work in regions sensitive to the red, together with the low limit of the radiations recorded, make this method practically useless for experimental work.

Following Langley, a series of most valuable investigations has been carried out by Rubens, Nichols, Aschkinass, Wood, von Baeyer and others working in this region. By means of the radiometer, the method of "reststrahlen" and focal isolation the investigations in the infra-red have been gradually extended to about 343 μ or roughly one-third of a millimeter.

Experimental Apparatus for Infra-red Transmission of Glass.

The refinements and the extended ranges of wavelength obtainable by several of the methods just mentioned are not necessary in investigating the absorption and transmission of the ocular media and of glass in the infra-red region. Tests by a considerable number of investigators show that the transmission of the eye media becomes very small after the region 3.5 μ is passed, while the transmission of ordinary glass drops, in general, to the order of 10 to 15 per cent. at 4.5 μ.

A quite satisfactory form of instrument for the examination of the transmission of glasses and similar media in the infra-red region is the Hilger infra-red spectrometer. The essentials of this instrument are shown diagrammatically in Figure 7. In Figure 8 is given a photographic reproduction of this instrument.

Radiation from a suitable source is allowed to pass through a narrow slit S of the order of magnitude of one-hundredth of an inch in width. The radiation from S is received by a concave mirror K by which it is collimated. It then falls on the rock-salt prism P, by which it is dispersed and received on the plane mirror M and reflected to the concave mirror R. This concave mirror R then focusses the radiation on the slit T, behind which is mounted a Hilger bismuth-silver thermopile which acts as the receiving instrument. The mirrors are made of nickeled steel. The slit at T is of the order of one one-hundredth inch in width. The prism P and the mirror M are mounted on a table which can be rotated around a vertical axis by means of a fine screw which is attached to a calibrated drumhead. From this drumhead the wavelength used for experimental purposes can be read directly. By this rotation of the prism P any desired part of

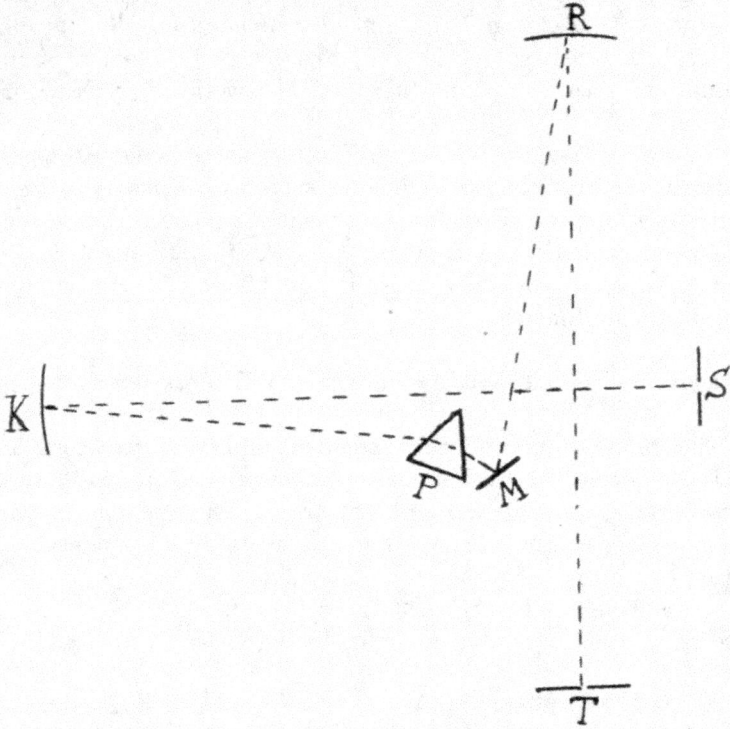

Fig. 7—Essentials of construction of the Hilger infra-red spectrometer.

the spectrum can be made to fall upon the slit T. The thermopile and the whole instrument must be carefully protected from external radiation.

The thermopile serves as the receiving instrument in this type of instrument. Nobili devised what he called a "pile" or a form of thermo-electric battery in which there are a large number of elements in a very small space. For this purpose he joined the couples of bismuth and antimony in such a manner that, after having formed a series of five couples as shown in Figure 9 (B) the bismuth from b was soldered to the antimony of the second series similarly arranged; the last bismuth of this to the antimony of the third and so on. The whole pile thus consisted of a number of bismuth-antimony couples. The couples can be insulated from each other by means of small paper bands covered with varnish and are then enclosed in a suitable frame P (Figure 9A) so that the only solderings appear at the two ends of the pile. Two small binding posts, m and n, insulated in an ivory ring, communicate in the interior, one with the antimony, representing the positive pole, and the other with the last bismuth, representing the

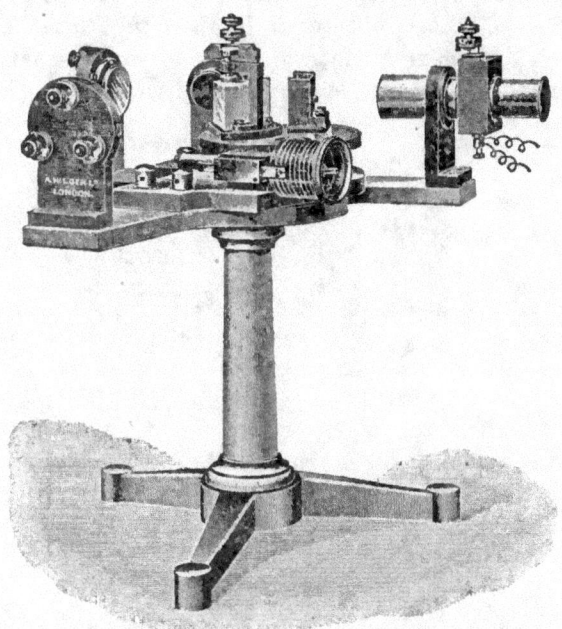

Fig. 8—The Hilger infra-red spectrometer.

negative pole. These terminal points then connect with a galvanometer: this instrument detects the thermo-electric current. The action of the thermopile depends upon the principle that if one set of junctions is at a higher temperature than the second set an electric current is produced. If these thermo-couples are made of the proper elements and are connected in the circuit of a sensitive galvanometer, extremely small fractions of a degree rise in temperature can be detected. The electric current arises in every case, however, because of the difference in temperature of the two faces of the thermo-junctions.

A sufficiently sensitive galvanometer is used as the instrument for the detection of the current. The strength of the current is proportional to the mirror deflection. This deflection can be measured by means of a lamp and scale.

For investigations in the infra-red transmission of glass it is found that the Nernst glower is a very satisfactory source. The distribution of energy in wavelengths of the emission of a Nernst glower varies considerably with the temperature. The radiation from such a glower is characterized by two maxima at about 2.5 μ and 5.5 to 6 μ. At low temperatures (2 watts to 7 watts) the latter maximum (5.5 μ) is the more prominent. As the temperature is raised (11 or more watts) the maximum of the energy distribution appears in the region of about 2 μ (See paper by Coblentz, *Bulletin of the Bureau of Standards*, Vol. 4, 1907.). This region from 0.7 μ to 3 μ is that which is of interest from the standpoint of transmission and absorption in the eye

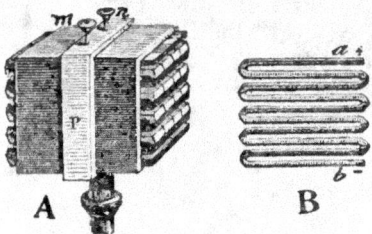

Fig. 9—The thermopile.

media and in various kinds of glass. The Nernst glower is, therefore, used in such experimental work with the maximum energy in the region of 1.5 to 2 μ.

Spectra of Illuminants.

The spectral distribution of energy in the radiation from different illuminants is of great importance in the consideration of color. This variation in the spectral character of illuminants is due to the temperature and composition of the radiating body and also to the state in which it exists when giving out luminous energy. A gaseous body gives out only certain definite rays and the spectrum is said to be a *line* spectrum. Quite often these spectral lines are crowded together in such a manner as to give to the spectrum a fluted or banded appearance. This is known as a *band* spectrum. Also, the constancy of the spectrum lines given by any substance (element) in gaseous form is a striking feature. For example, the visible spectrum of sodium consists of a double line (5890 t. m. and 5896 t. m.) and whenever this double line is found in a spectrum it is certain that sodium is

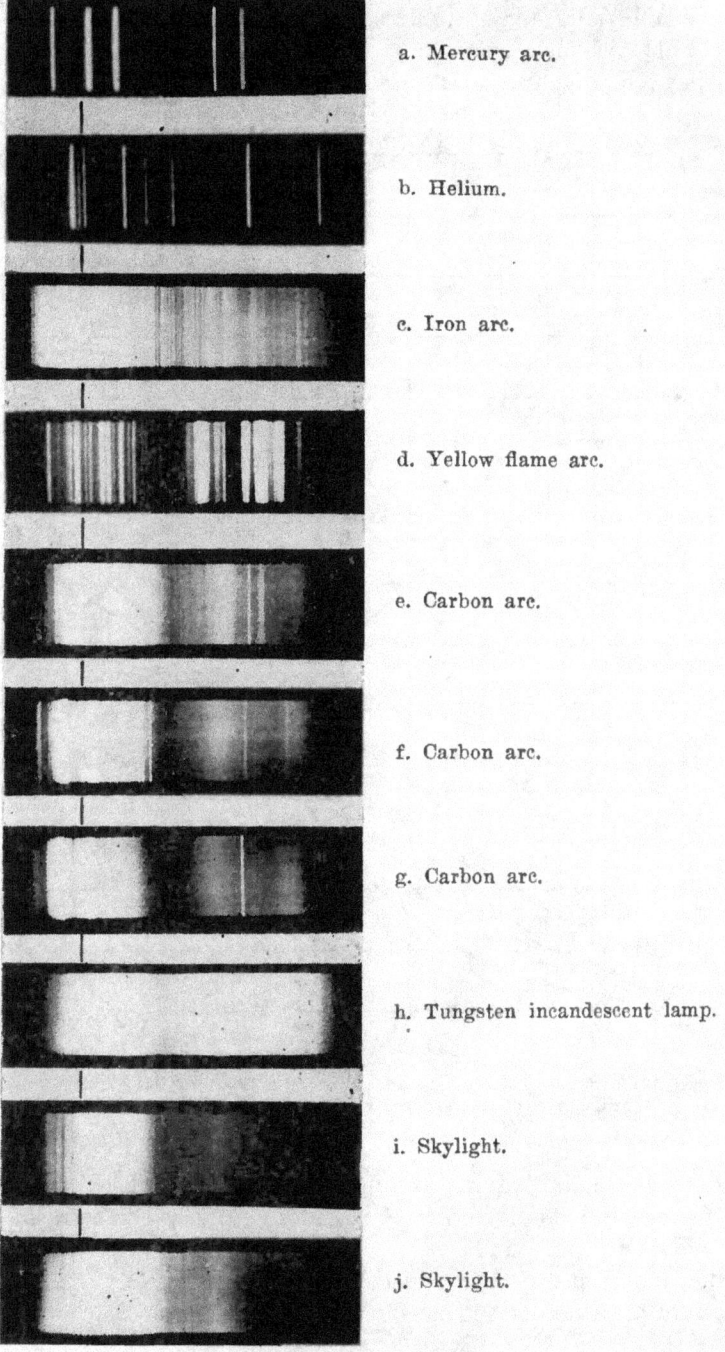

a. Mercury arc.

b. Helium.

c. Iron arc.

d. Yellow flame arc.

e. Carbon arc.

f. Carbon arc.

g. Carbon arc.

h. Tungsten incandescent lamp.

i. Skylight.

j. Skylight.

Fig. 10—Representative spectra. (Courtesy of M. Luckiesh.)

present in the radiating substances. This constancy of spectra forms a basis of analysis more sensitive than the most accurate chemical tests. The element helium was discovered by means of spectroscopy some-

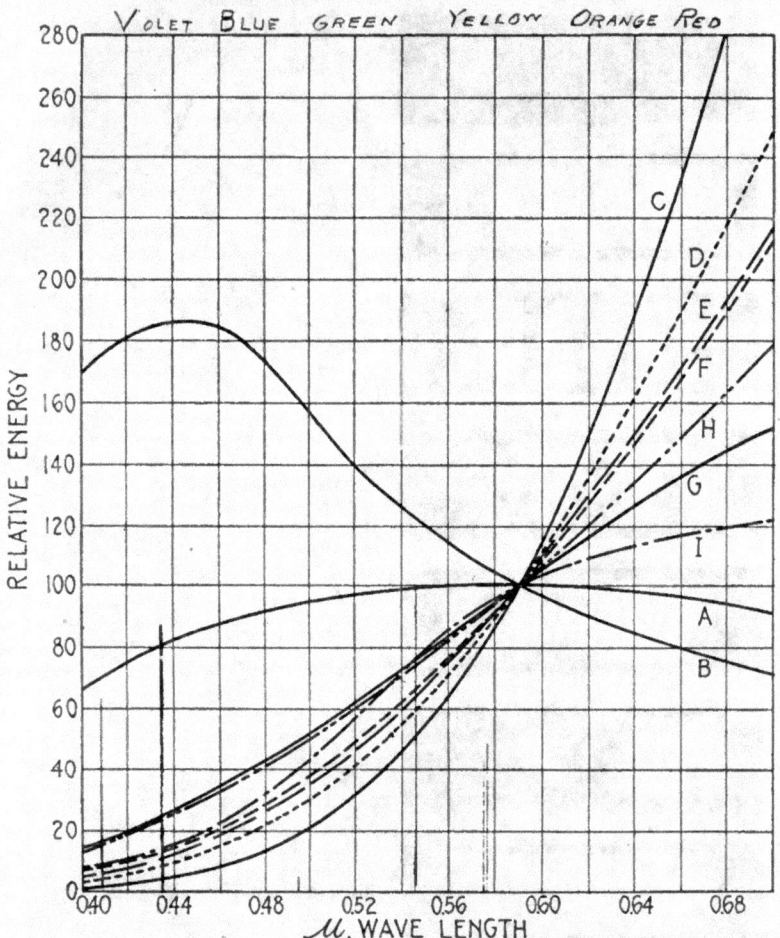

Fig. 11—Distribution of energy in the visible spectrum of various illuminants. Significance of letters on curves given in Table II.
(Courtesy of M. Luckiesh.)

time before it was terrestrially found. The vacuum tube, the electric spark, the arc and the flame are of use in studying the spectra of elements and their compounds.

A continuous spectrum is emitted by an incandescent solid. The spectrum of an incandescent electric lamp, for example, is continuous.

The energy of the electric current running through such a filament is converted into radiant energy. The continuous spectrum is, as its name signifies, the antithesis of the line spectrum: or it may be considered as an infinitely numbered line spectrum. There are no breaks or apertures in the emission. Sometimes both a line and a continuous spectrum are emitted by an illuminant. Such a condition exists in the ordinary electric carbon arc. The center of the arc is an incandescent solid and therefore emits visible rays of all wavelengths; the incandescent gas of the arc between the electrodes emits a line spectrum which depends as to its appearance upon the surrounding medium and the character of the carbon electrodes. In Figure 10 are shown several representative spectra photographed by means of a sensitive spectrograph using Cramer spectrum plates which were made specially sensitive to the visible rays (Luckiesh: *Color and Its Applications*, page 17). The reproduced spectrograms contain line spectra, banded spectra and continuous spectra. It will be seen that the two gases, mercury and helium, emit line spectra. The arcs emit both continuous and line spectra. The relative prominence of the line spectra depends upon the relative intensities of the radiation from the arc as compared with that from the solid electrodes. For instance, the line spectrum is much more prominent in the yellow flame arc than in the ordinary carbon arc. As is well known, the arc vapor contributes a much greater proportion of the light in the former than in the latter illuminating source. The line spectrum of carbon is subject to changes because of the character and amounts of impurities which may be present in the carbons. The three spectra of the carbon arc given in Fig. 10 were taken within a few minutes' time and show these variations. (The apparent absorption in the green region in all these photographs is due to lack of sensitiveness of the plates used in the green-blue region). The tungsten filament, *h*, is seen to emit a continuous spectrum. Two spectrograms of light from the sky are shown in *i* and *j* and bring out (perhaps rather poorly) the presence of narrow black absorption lines. The solar spectrum is of interest particularly because of the fact that it is a continuous band crossed by many fine dark lines. These lines were discovered in all probability by Wollaston in 1802 but were studied with better instruments by Fraunhofer in 1814 and are consequently known as Fraunhofer lines. These absorption lines are due to the removal of the corresponding radiations by the vapors in the solar atmosphere. The chief Fraunhofer lines with their wavelengths, colors and sources are given in Table I.

TABLE I.

Principal Fraunhofer Lines.

Line.	Wave-length.	Color.	Source.
A	0.7594μ	Red	Oxygen in atmosphere
a	0.7185	Red	Water vapor
B	0.6876	Red	Oxygen in atmosphere
C	0.6563	Red	Hydrogen in sun
D_1	0.5896	Yellow	Sodium in sun
D_2	0.5890	Yellow	Sodium in sun
E	0.5270	Green	Calcium in sun
b_1	0.5184	Green	Magnesium in sun
b_2	0.5173	Green	Magnesium in sun
b_4	0.5168	Green	Magnesium in sun
F	0.4861	Blue	Hydrogen in sun
G	0.4308	Violet	Calcium in sun
H	0.3969	Violet	Calcium in sun
K	0.3934	Violet	Calcium in sun

Figure 11 gives curves showing the spectral distribution of energy in the visible region for various illuminants. These data were obtained chiefly by Hyde, Ives, Cady and Luckiesh working in the Nela Research Laboratory. Table II gives the numerical data as well as the significance of the letters attached to the different curves (Luckiesh,

TABLE II.

	A	B	C	D	E	F	G	H	I
Wave length	Block Body at 5000° Absolute (Noon Sunlight)	Blue Sky	Hefner lamp	Carbon incandescent lamp 3.1 w. p. m. h. c.	Acetylene	Tungsten incandescent lamp 1.25 w. p. m. h. c.	Tungsten incandescent lamp 0.5 w. p. m. h. c.	D. C. arc	Welsbach gas mantle
0.41 μ	72.0	177	1.9	4	5.5	....	16.5	....	....
.43	79.0	185	3.5	7	9.6	....	22.5	21.8	....
.45	84.3	187	6	12	15	16.7	30	29	17.5
.47	91.0	180	10.5	18	21.9	23.5	38	37	26.4
.49	92.5	162	16.3	25.5	30.3	32.7	47	45.5	38.3
.51	96.0	146	25.5	34.5	40	42.6	56.5	55	51
.53	98.0	132	37.5	47	52	54.9	67	65.5	64
.55	99.0	120	53.2	62	66.5	68.6	78	76	78
.57	100.0	108	74.5	79	82	83.4	88	88	90
.59	100.0	100	100	100	100	100	100	100	100
.61	100.0	93	130	123	118	117	111	113.5	107
.63	98.5	87	168	148	139	136	121	127	111
.65	97.1	82	210	176	160	157	131	142	114
.67	95.5	77	260	204	182	179	140	156	119
.69	93.5	72.5	320	204	205	202	148	170	120

Color and Its Applications, page 21). It will be noted that all curves are plotted in such a manner that the relative energy of wave-length 0.59 μ (approximately sodium D) equals one hundred. This method of plotting gives the relative distribution of energy for approximately the same amounts of total light sensation. All of these curves show

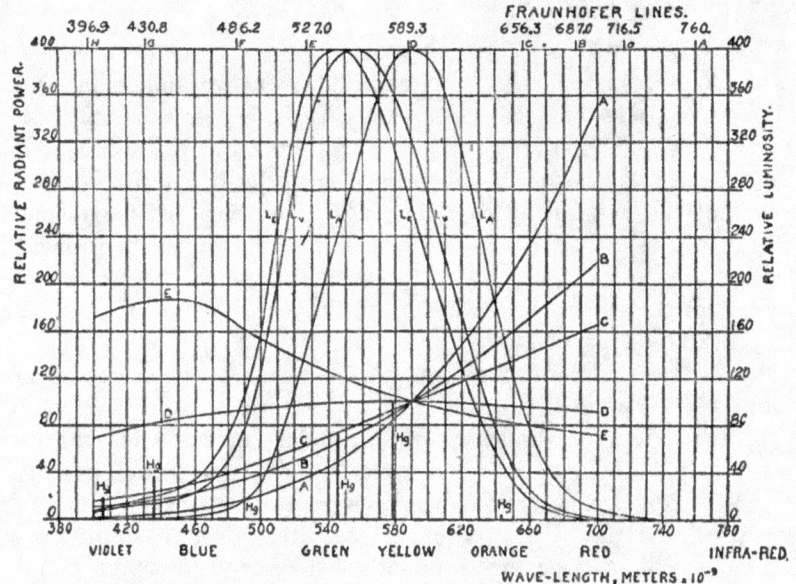

Fig. 12—Relative spectral distribution of radiant power in various sources: *A*=Hefner lamp, H. E. Ives, Trans. I. E. S., 5, p. 208, 1910; *B*=acetylene flame, Coblentz and Emerson, B. S., 13, p. 363, 1916; *C*=tungsten (gas) incandescent lamp (No. 1717) at 15.6 lumens per watt, data by Coblentz; *D*=black body at 5,000° absolute—approximately sunlight, computed from Planck equation; *E*= blue sky, H. E. Ives, Trans. I. E. S., 5, p. 208, 1910; *Hg*=Heraeus quartz-mercury lamp, 81 watts, W. W. Coblentz, B. S., 9, p. 97, 1913.

Relative visibility: *Lv*=Relative visibility curve for the average human eye (or luminosity of a source having constant radiant power throughout the visible spectrum), Coblentz and Emerson, B. S., 14, p. 192, 1917.

Relative luminosity=relative radiant power times relative visibility: *LA*= luminosity of Hefner lamp; *LE*=luminosity of blue sky.

(From Technologic Paper No. 119, 1919. Permission of The Bureau of Standards.)

that all artificial light sources, such as the 'Welsbach, acetylene and tungsten incandescent lamps are relatively rich in longer visible wave-lengths and decidedly deficient in terms of percentages in the extreme blue and violet.

Figure 12 is taken from a recent paper by Gibson and McNicholas (The Ultra-violet and Visible Transmission of Eye-Protective Glasses, *Bulletin of Bureau of Standards,* June, 1919) and gives data by Ives and by Coblentz. The descriptive matter accompanying the diagram makes clear the significance of the curves.

CHAPTER III. TRANSMISSION AND ABSORPTION OF GLASS FOR ULTRA-VIOLET, VISIBLE AND INFRA-RED RADIATION.

The Ultra-violet and the Visible.

It is a well known fact that ultra-violet light (light of wave-length less than 3900 t. m. roughly) may exert harmful physiological effects on the eye and skin, but just how much of this deleterious action is to be ascribed to general energy radiation and how much to specific radiation is a matter that has by no means been settled. It is quite generally agreed that the extreme ultra-violet rays, i. e., those of wave-length 3000 t. m. or less, cause injury when in sufficient quantity or intensity. There are those who claim, and with considerable evidence in support thereof, that the rays between 3000 and 3600 t. m. also cause injury. Nutting (*Bureau of Standards*, Circular No. 28) believes that the 3650 t. m. of the mercury arc contributes about 80 per cent. of the "fatigue effect" when this arc is used as a source of light. Whatever may be the extent of the extremely harmful regions and whatever opinions may be held as to what radiations are or are not harmful, there are still many industrial processes requiring special protection of the eyes, and the excessive amount of ultra-violet or of infra-red radiation may be and generally is one of the very important factors. It is therefore a matter of importance to know much of the physiological and pathological effects of radiation: these we shall consider in another part. It is likewise of great importance that we should know how much of a specified radiation gets through a given sample of glass or other absorbing medium.

In 1889 Widmark (*Skand. Arch.* I, 264) made experiments on the subject of the effects of ultra-violet light on the eyes of laboratory animals and reproduced the stages of electric ophthalmia in rabbits' eyes. Perhaps stimulated by the possibilities of the protection of the eyes suggested by Widmark's experiments, Schulek (*Ungar. Beitr. z. Augenheilk.* I, 101, 1895 and 2, 1899) first studied the means of protecting the eyes against ultra-violet rays and found that certain liquids had the highest absorptive powers of the transparent media investigated. These liquids absorbed all rays below 3960 t. m. He suggested that these solutions should be enclosed in flat, oval-shaped glass chambers made to fit the eyes and to protect them from injuries due to the ultra-violet radiations.

Stearkle (*Arch. f. Augenheilk.* 50, 1904), Vogt (*Arch. f. Augen-*

heilk. 59, 1907) and Hallauer (*Vers. d. Ophth. Ges. Heidelberg,* 1907) studied the absorptive properties of blue uviol, yellowish and smoky-gray glasses. The last named worker produced by a secret process the so called Hallauerglas. Following a like study Fieuzal (*Bull. de la clin. nat. oph.* No. 3, 1885) produced a glass known as Fieuzalglass. Also a yellow-green glass patented under the name of Enixanthosglas was offered, as well as a variety of modifications.

In 1907 Schanz and Stockhausen (*Klin. Monatsbl. f. Augenheilk.* 1907) after finding that electric ophthalmia could be produced through 18 mm. of common glass, began to study the problem of manufacturing a colorless glass of high ultra-violet absorptive power. In 1909 they produced and patented a glass of higher absorptive power than hard flint and named it Euphosglas. This glass has a light, yellowish-green tinge and fluoresces in ultra-violet light.

In 1909 Birch-Hirschfeld (*Zeitschr. f. Augenheilk.* 21, 1909) studied photometrically the absorptive power of various glasses with considerable accuracy. At about the same time Vogt (*Arch. f. Augenheilk.* 59, 1907) compared a new and very hard flint glass produced by Schott with his absorptive solutions and found that it had about the same absorptive efficiency, beginning at 4050 t. m. and giving practically complete absorption below 3960 t. m. Hallauer (*Arch. f. Augenheilk.* 54, 1909) measured photometrically the absorption powers of the various protective glasses then available.

As the number of kinds of glass for cutting out various wavelengths became greater and more available, the question arose as to what spectral range constituted the best illumination. Voege (*Electro. Zeits.* No. 33, 1908) maintained that the light from the clouds or clear sky had been for ages a normal illumination for the eyes and that it contained a considerable amount of ultra-violet light as low as 3000 t. m. Hertel and Henker (*Arch. f. Ophthal.* 73, 1910) carried out a very elaborate set of experiments in the Zeiss Laboratory, Jena, and came out in support of Voege. These experimenters believed that the best glass is one that will reduce the spectrum of the particular light to which the eyes are exposed to the closest possible approximation to the spectrum of cloud and sky lights. For observation of the strongest arc lights at close range they considered the neutralglas F 3815 of Schott to be best, claiming that this glass, in layers thinner than any other glass, may be used to observe directly a bare 20 amperes arc light at a foot and a half without injury, since the spectrum is about the same as that of cloud light minus the ultra-violet portion. Schanz and Stockhausen (*Arch. f. Ophth.* 75, 1919) criticized this work, particularly on the basis that skylight or cloud light cannot be taken as the ideal light. They furthermore cite the work of Handmann (*Monatsbl. f. Augenh.* 47, 1909) showing that a very large group of cataracts began in the quadrant of the lens most exposed during life to the light of the sky. It may be stated, in brief, that during the past ten years these various discussions and opinions have caused the number of protective eye glasses for general wear or specific purposes to be multiplied and attention to be paid to glasses affecting the ultra-violet, the visible and the infra-red.

No survey of the development of protection glasses would be complete without mention of the paper on The Preparation of Eye-Preserving Glass for Spectacles (*Trans. of Roy. Soc.*, 1913) delivered by Sir William Crookes before the Royal Society on November 13th, 1913. Crookes was engaged from 1909 to 1913 in connection with the Glass Workers Cataract Committee of the Royal Society and experimented on the effect of adding various metallic oxides to the constituents of glass in order to cut off the ultra-violet and the infra-red rays. The main object of the researches was to prepare a glass which would cut off the rays from highly heated molten glass which apparently damaged the eyes of workmen. Photo-spectrographic and other examinations were made of the radiation emitted from the molten glass under working conditions. Sir Wm. Crookes, in the paper referred to, writes: "Taking the ordinary limit of visibility to lie between 3900 t. m. and 7600 t. m. it is seen that with an exposure of three hours to the highest heats the strength of impression does not extend much into the ultra-violet. The heat rays are very strong and if injury to the eye is caused by exposure to radiation from the molten glass, a protective glass should be opaque to infra-red rays. These being present in the radiation from molten glass in far greater abundance than the ultra-violet rays, the inference is that it is to the heat rays rather than to the ultra-violet rays that glass workers' cataract is to be ascribed. It is, however, certain that exposure to excess of ultra-violet light also injuriously affects the eye. That the ultra-violet rays act on the deeper-seated portions of the eye is shown by the intense fluorescence of the crystalline lens induced by these rays. Besides the invisible rays at each end of the spectrum, the purely luminous rays, if present in abnormal intensity, are found to damage the eye. It, therefore, would be an advantage if in addition the obscuring glass for the spectacles were to be of a neutral or gray tint."

In discussing the results of his experimental work Crookes says that the first necessity is to find a glass which will cut off as much as possible of the heat radiation. "For ordinary use," he writes, "when no special protection against heat radiation is needed, the choice will

rest on whether the ultra-violet or the luminous are most to be guarded
against; or whether the two together are to be toned down.'' His
experimental work gave glasses which were very effective in cutting
out wavelengths shorter than 3700 t. m. The colors of these glasses
were pale-green, yellow and neutral. Likewise glasses of much transparency were produced which transmitted from 99.5 to 70 per cent. of
the incident light. The choice between this range of glasses would
depend on the conditions required. Special glasses were devised which
are ''restful to the eyes in the glare of the sun on chalk cliffs, expanses
of snow, or reflected from the sea.   *   *   *   Moreover, they have the
advantage of cutting off practically all of the ultra-violet rays and also
a considerable amount of the heat radiation.''

While a great deal of work has been done on ultra-violet light, both
in Europe and in America, very few *quantitative* investigations have
been made. Bell and Luckiesh were two of the first experimenters to
make quantitative investigations along these lines. Bell (*Electrical
World,* April, 1912 and *Amer. Acad. Proc.* 46, April, 1911, etc.), by
means of a thermopile and sensitive galvanometer, has obtained valuable data on the ultra-violet component of artificial light sources.
Luckiesh (*Electrical World,* June, 1912, *Illuminating Eng. Soc.,* April,
1914, *Elect. World,* May 24, 1913, etc.) used a photographic method
to obtain transmission curves of various kinds of glass.

Ham, Fehr and Bitner, using a photographic null method for measuring absorption in the ultra-violet, published their results (*Journal
of the Franklin Institute,* Sept., 1914) upon the transmission and minimum lines of various kinds of glass.

In 1918 Coblentz and Emerson (*Technologic Papers of the Bureau
of Standards,* No. 93, 1918) issued a paper on the subject of Glasses
for Protecting the Eyes from Injurious Radiations. This paper deals
quite largely with the visible region beyond 0.5μ and with the absorption of glasses in the red and infra-red regions. Coblentz and Emerson conclude: ''For protecting the eye from ultra-violet light, black,
amber, green, greenish-yellow, and red glasses are efficient. Spectacles
made of white crown glass afford some protection from the extreme
ultra-violet rays which come from mercury-in-quartz lamps and from
electric arcs between iron, copper or carbon. The vapors from these
arcs emit but little infra-red radiation in comparison with the amount
emitted in the visible and in the ultra-violet.''

In June 1919 Gibson and McNicholas issued a paper (*Technologic
Papers of the Bureau of Standards,* No. 119, 1919) on the Ultra-violet
and Visible Transmission of Eye-Protective Glasses, in which they
report the results of a long series of careful spectrophotometric obser-

vations for different wavelengths upon various eye-protective media. A logarithmic relation connects the transmittance and the thickness of glass, thereby enabling a direct comparison of the transmissions and absorptions of various kinds of glass of different thicknesses. A considerable number of their results in the form of curves appear in other portions of this essay.

Let us consider some of these experimental methods and the results of various investigators somewhat in detail. There are various means and methods of studying the transparency of media for the ultraviolet. Photography is, without doubt, the most readily applicable. The radiometer, thermopile and bolometer could be used, but temperature changes, air currents, magnetic disturbances and general inconveniences of such methods bar them out as useful in such investigations as are now being discussed. Photography offers several distinct advantages: among these may be mentioned: (1) Less adjustment than is required in any other method, (2) extremely faint lines can be detected and measured and (3) the photographic plate gives a permanent record of the test. However, be it said that when the transmission is to be accurately determined the photographic method is a very tedious procedure. The photographic action is determined by the density of the plate. In plotting the density of the plate against the logarithm of the illustration a straight line relation is found over a certain range of illumination. By no means, however, is the density of the silver deposit proportional to the logarithm of the intensity of radiation throughout any extremely wide range of illumination. Furthermore, rays of various wavelengths show different relations between density of the silver deposit and the illumination.

Ham, Fehr and Bitner (*Journal of the Franklin Inst.*, page 299, 1914) used the null photographic method of determining transmissions. By making several exposures on the same plates, various sources of error such as changes in temperature, character of plate, and so forth, could be eliminated and there would be practically no errors introduced due to the emulsion and the development. By making several exposures with various reduced intensities of the incident beam a very close match could be obtained between two adjacent images of the same spectral line and a fairly close estimate of the absorption obtained for that particular wavelength. For example, if the effect produced by the original beam of light after passing through the medium were the same as the effect produced by the beam when striking the plate after a reduction of 25 per cent. in intensity, the absorption of the medium would be 25 per cent. of that particular wavelength. The experimental problem, therefore, consisted of two parts: (1) the determination of the equality of the densities of the two adjacent images on the photographic plate and (2) the reduction of the incident beam of light by a known amount.

Figure 13 (Ham, Fehr and Bitner, *Journal of the Franklin Institute*, Sept., 1914) shows how important it is in such work to increase this time of exposure until no more lines appear on the spectrogram. With the apparatus used it was found that an exposure of 120 seconds was sufficiently long to bring out the line of minimum wave-

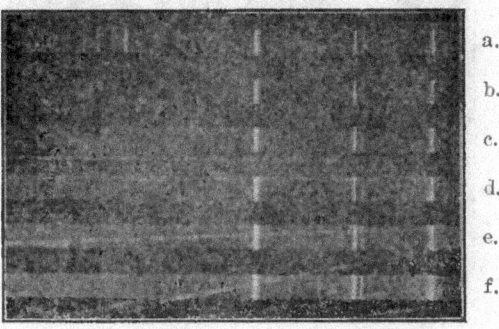

Fig. 13—Transmission of clear glass. (After Ham, Fehr and Bitner, *Jour. of Franklin Inst.*, 1914.)

a. Quartz, 10 seconds.
b. "American" clear glass, 10 seconds.
c. "American" clear glass, 20 seconds.
d. "American" clear glass, 30 seconds.
e. "American" clear glass, 60 seconds.
f. "American" clear glass, 120 seconds.

length in all the glasses tested, whether high or low in the transmission of visible light.

Figures 14, 15, 16 and 17, taken from the same paper, were made with exposures of 120 seconds each. The percentage transmission was obtained by the use of a 1.25 watt per candle tungsten lamp and a flicker photometer. The authors say: "In looking over the data obtained in regard to the minimum wavelength transmitted, some very interesting results may be noted. For example, the faint pink of No. 4 transmits as much ultra-violet as the light blue of No. 20. Since both glasses have about equal transmissions for visible light, it was to be expected that the one nearer the red end of the spectrum would cut off more ultra-violet, but this case clearly shows that no dependence may be placed on the color of the glass. Again, Nos. 7 and 8, of very nearly the same shade of yellow, show widely different degrees of transmission of ultra-violet, the 'Noviol' not transmitting even all of the visible wavelengths while the other yellow glass transmits as far down as the 3150 t. m. line. Euphos glass No. 11 cuts off the ultraviolet very sharply at 4050 t. m. but Nultra glass appears to be somewhat better from a practical standpoint for it barely transmits the 3650 line (less than 1 per cent. by actual test) and absorbs only 15 per cent. of the visible light. The most remarkable glass of all is the orange-yellow of No. 5 (figure 18) which appears to transmit selectively be-

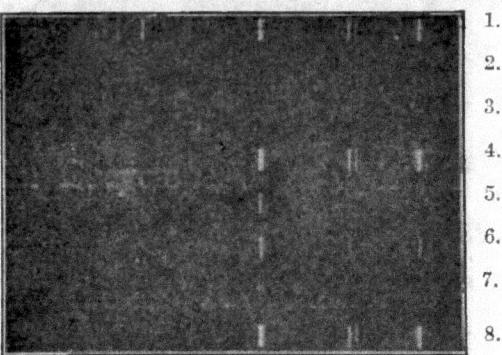

1.
2.
3.
4.
5.
6.
7.
8.

Fig. 14—Minimum lines transmitted by various glasses. (After Ham, Fehr and Bitner, *Jour. Franklin Inst.*, 1914.)

No.	Absorbing Medium	Per Cent. Transmission for Tungsten Light at 1.25 w.p.c.	Minimum Line $\mu\mu$
1.	Quartz	....	...
2.	Very deep red glass	4.1	...
3.	Red glass	17.8	...
4.	Faint pink	50.1	313
5.	Orange yellow	38.3	334
6.	Yellow	54.3	334
7.	Yellow ("Noviol")	78.0	546
8.	Very light yellow	78.8	313

tween 3340 and 4050 t. m., although its greatest transmission is at the other end of the spectrum."

One criticism of this work of Ham and his colleagues is that the mercury arc spectrum, with its comparatively few lines, was used as a light source. Hence the minimum transmission may not be the minimum line recorded. The remedy lies in the use of a continuous spectrum, or as nearly continuous as is obtainable. A method of underwater spark has been already referred to as being of great use in such work.

Smith and Sheard (*Journal of the Optical Society of America,* page 26, 1919) made use of a condensed spark across two electrodes, one made of iron, the other of an alloy, of cadmium, aluminium, magnesium and zinc. Figure 18 gives the photographic results of the minimum transmissions of the various samples of glass specified at the side. A Fery quartz spectrograph was used. "The times of exposure were made nearly the same throughout the experiment. The intensity of spark, however, fluctuated so much that it is not possible to make comparisons concerning the amount absorbed by the different thicknesses.

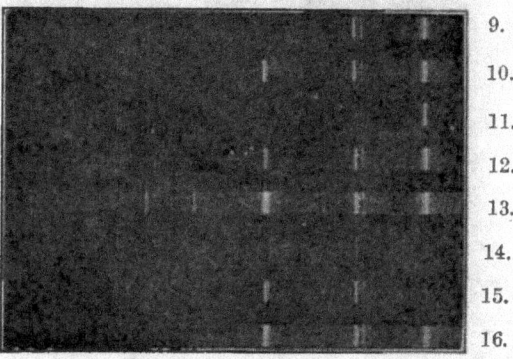

Fig. 15.—Minimum lines transmitted by various glasses. (After Ham, Fehr and Bitner, *Jour. Franklin Inst.*, 1914.)

No.	Absorbing Medium	Per Cent. Transmission for Tungsten Light at 1.25 w.p.c.	Minimum Line $\mu\mu$
9.	Faint yellow ("Nultra")	85.0	365
10.	Faint yellow	84.3	313
11.	Yellow green ("Euphos")	72.0	405
12.	Yellow green	71.6	365
13.	Faint yellow green	82.5	302
14.	Very deep green	2.6	405
15.	Dark green	5.0	365
16.	Green	52.0	334

* * * It was the purpose of this part of the experiment to show only limits to which these glasses transmit radiations in the ultra-violet end of the spectrum for fairly long exposures. From the results obtained it is seen that the amethyst and the blue glasses transmit farthest into the ultra-violet. They seem to absorb all wavelengths beyond 3091 t. m. On the other hand the deeper-colored Noviol transmits least far into the ultra-violet. It apparently absorbs everything beyond about 5000 t. m." No quantitative measurements are attached to these spectrograms but they do point out the fact that ambers of various kinds (called by such names as oliveye, luxfel,

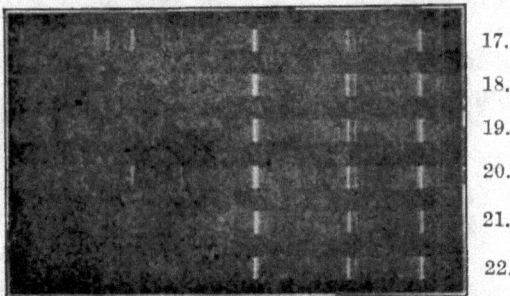

Fig. 16.—Minimum lines transmitted by various glasses. (After Ham, Fehr and Bitner, *Jour. Franklin Inst.*, 1914.)

No.	Absorbing Medium	Per Cent. Transmission for Tungsten Light at 1.25 w.p.c.	Minimum Line $\mu\mu$
17.	Quartz	...	...
18.	Dark blue violet glass.........................	2.5	334
19.	Blue	8.3	334
20.	Light blue ("Tungsten")	46.1	302
21.	Reddish purple	23.2	334
22.	Dark flesh color..............................	44.3	334

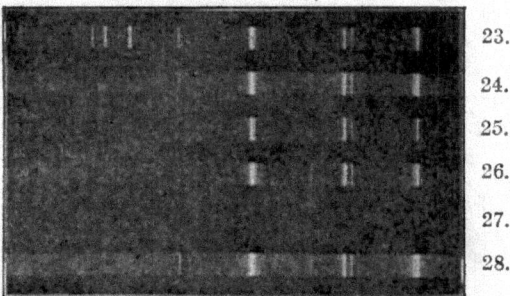

Fig. 17.—Minimum transmission of various glasses. (After Ham, Fehr and Bitner, *Jour. Franklin Inst.*, 1914.)

No.	Absorbing Medium	Per Cent. Transmission for Tungsten Light at 1.25 w.p.c.	Minimum Line $\mu\mu$
23.	Quartz	...	...
24.	Light flesh color	73.8	334
25.	Dark gray	15.3	334
26.	Medium gray	36.8	334
27.	Ground glass	50.5	365
28.	Clear glass ("American")	90.2	313

	Thickness of Glass
Condensed Spark	......
Euphos	5.0 mm
Pfund	2.6 mm
Crookes A	3.2 mm
Crookes B	2.8 mm
Noviol b.	4.9 mm
Noviol a.	4.0 mm
Smoke No. 0........	2.3 mm
Smoke No. 1........	2.3 mm
Smoke No. 2........	2.3 mm
Luxfel	2.4 mm
Oliveye	2.5 mm
Amethyst No. 1......	2.6 mm
Amethyst No. 2......	4.8 mm
Amethyst No. 3......	3.3 mm
Resistal	3.4 mm
Blue No. 0..........	2.5 mm
Blue No. 1..........	2.5 mm
Blue No. 2..........	2.5 mm
First Amber	4.1 mm
Light Amber	2.5 mm
Medium Amber	4.0 mm
Dark Amber	4.3 mm
Dark Amber No. 2...	2.3 mm
Light Amber No. 2...	2.4 mm
Nactic a.	3.4 mm
Nactic 21	3.3 mm
Nactic 22	3.0 mm
Nactic 23	3.2 mm
Nactic 24	3.5 mm

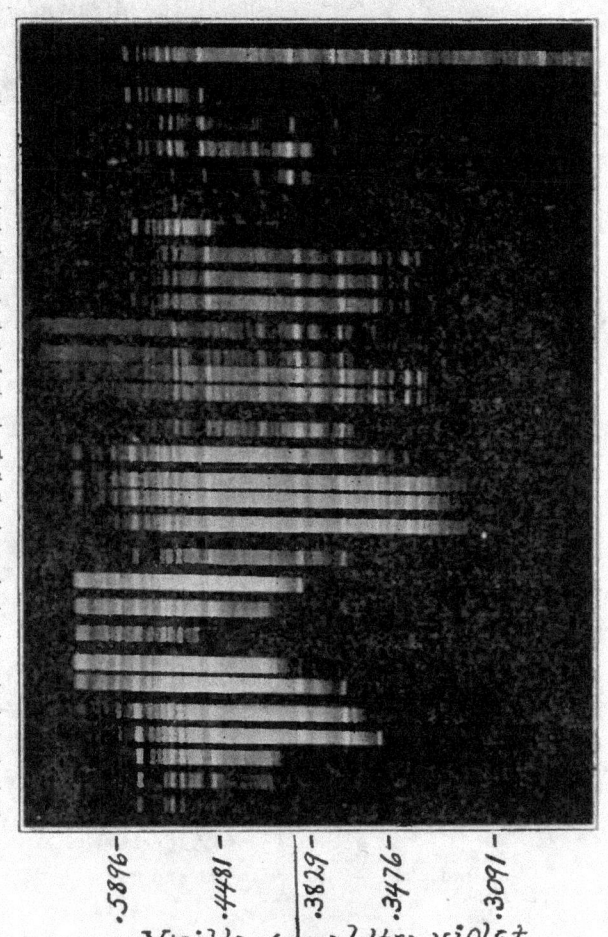

Fig. 18.—Transmission of various ophthalmic glasses in the visible and ultra-violet. Spectrographic record. (Permission of the *Journal of the Optical Society of America.*)

ambers and nactics, etc.) are entirely different in their transmission limits and hence in the amounts of various wavelength energy which they will transmit. The family of spectrograms labelled "Nactics" at the bottom of Figure 18 illustrates this point very nicely. Neither, in turn, can the relative limits of absorption be estimated in many cases by gradation of color of glass. In other words, coloring chemicals may often be added to glass without appreciably affecting the limit of transmission, as instanced by Crookes A and B shades, in which the limit of transmission in the ultra-violet is virtually the same. And again, various samples of glass, all possessing the same color as judged by a matching of samples laid on a sheet of white paper, may vary considerably in their limits of transmission.

In 1914 Luckiesh presented a paper before the Illuminating Engineering Society (*Transactions of the Illuminating Eng. Soc.*, Vol. 9, 1914) on "Glasses for Protecting the Eyes." Luckiesh adopted the procedure of using the light from a quartz mercury arc reflected from a magnesia block. A series of exposures of equal length was made but with different known illuminations of the magnesia surface. Fol-

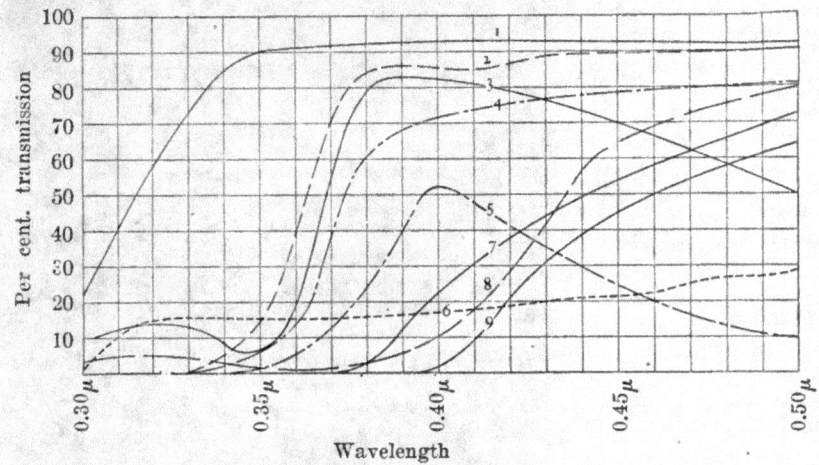

Fig. 19—Transmission of glasses in the region 3000 t. m. to 5000 t. m. (Courtesy of M. Luckiesh.)

1. Clear lead glass	4. Light amber	7. Medium amber
2. D smoke	5. 7 smoke	8. Euphos
3. Amethyst	6. 6 smoke	9. Akopos

lowing these, exposures of the same length were made through the media to be examined with known illumination. The photographs were measured for density on a Martin's polarization photometer and curves were plotted between density and illumination for each line. From these curves the corresponding intensities of illumination (*i. e.*, transmission) were read off for each line of each negative exposed through the specimens. By taking into account the relative illuminations of the magnesia block the transmission at each wavelength was readily obtained. Figure 19 shows the transmission curves of various glasses in the region of 0.3μ to 0.5μ (3000 to 5000 t. m.). The curves are numbered and the glasses giving these transmissions are tabulated above the curves. Luckiesh says: "It will be noted that the trans-

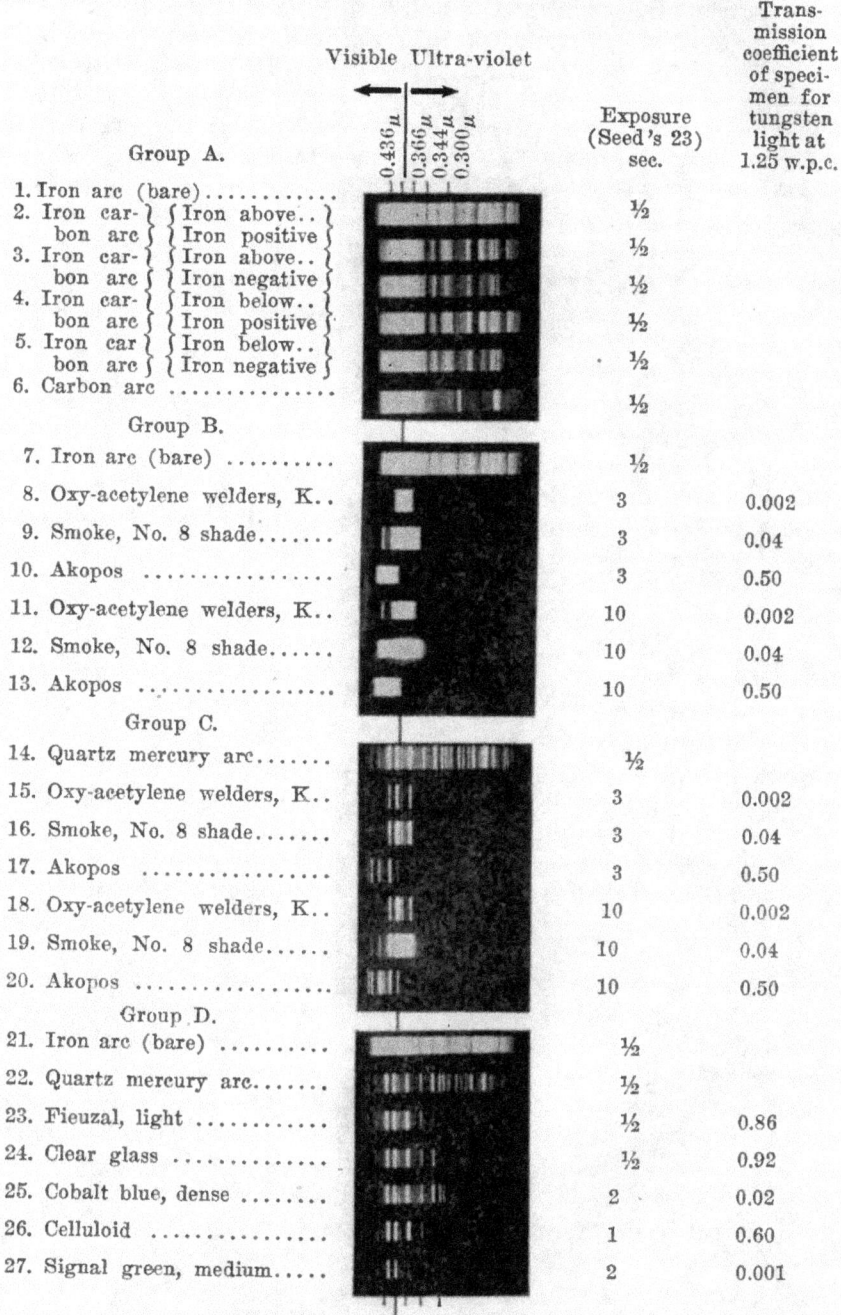

	Visible	Ultra-violet		Exposure (Seed's 23) sec.	Transmission coefficient of specimen for tungsten light at 1.25 w.p.c.

Group A.

1. Iron arc (bare).......... — ½
2. Iron carbon arc { Iron above.. / Iron positive } — ½
3. Iron carbon arc { Iron above.. / Iron negative } — ½
4. Iron carbon arc { Iron below.. / Iron positive } — ½
5. Iron carbon arc { Iron below.. / Iron negative } — ½
6. Carbon arc .... — ½

Group B.

7. Iron arc (bare) .... — ½
8. Oxy-acetylene welders, K.. — 3 — 0.002
9. Smoke, No. 8 shade....... — 3 — 0.04
10. Akopos .... — 3 — 0.50
11. Oxy-acetylene welders, K.. — 10 — 0.002
12. Smoke, No. 8 shade...... — 10 — 0.04
13. Akopos .... — 10 — 0.50

Group C.

14. Quartz mercury arc....... — ½
15. Oxy-acetylene welders, K.. — 3 — 0.002
16. Smoke, No. 8 shade....... — 3 — 0.04
17. Akopos .... — 3 — 0.50
18. Oxy-acetylene welders, K.. — 10 — 0.002
19. Smoke, No. 8 shade...... — 10 — 0.04
20. Akopos .... — 10 — 0.50

Group D.

21. Iron arc (bare) .... — ½
22. Quartz mercury arc....... — ½
23. Fieuzal, light .... — ½ — 0.86
24. Clear glass .... — ½ — 0.92
25. Cobalt blue, dense .... — 2 — 0.02
26. Celluloid .... — 1 — 0.60
27. Signal green, medium..... — 2 — 0.001

Fig. 20—Spectrograms of the transmission of various glasses in the ultraviolet. (Courtesy of M. Luckiesh.)

Visible | Ultra-violet

0.436μ 0.366μ 0.344μ 0.300μ

	Exposure (Lantern slide plate) sec.	Transmission coefficient of specimen for tungsten light at 1.25 w.p.c.
Group E.		
28. Quartz mercury arc......	½	
29. Clear glass, 1/16″ thick.....	½	0.92
30. Amber, medium, X.......	4	0.50
31. Kosma	1	0.83
32. Electric smoke	480	0+
33. Smoke X	3	0.015
Group F.		
34. Quartz mercury arc.......	½	
35. Amber, medium, X.......	15	0.50
36. Kosma	½	0.83
37. Smoke X	6	0.015
38. Euphos, 3/64″ thick	8	0.81
39. Thermoscopic	1	0.86
Group G.		
40. Quartz mercury arc	½	
41. Distilled water (1 cm.)...	½	0.93
42. Clear glass, 1/16″ thick....	½	0.92
43. Smoke A	1½	0.36
44. Smoke C	2½	0.20
45. Smoke A + C	7	0.07
Group H.		
46. Quartz mercury arc	½	
47. Euphos, 1/64″ thick.......	1	0.90
48. Amber, light shade.......	3	0.67
49. Amber, medium shade.....	6	0.45
50. Amber, medium, X.......	20	0.50
51. Smoke X	15	0.015

Fig. 21—Spectrograms of the transmission of various glasses in the ultraviolet. (Courtesy of M. Luckiesh.)

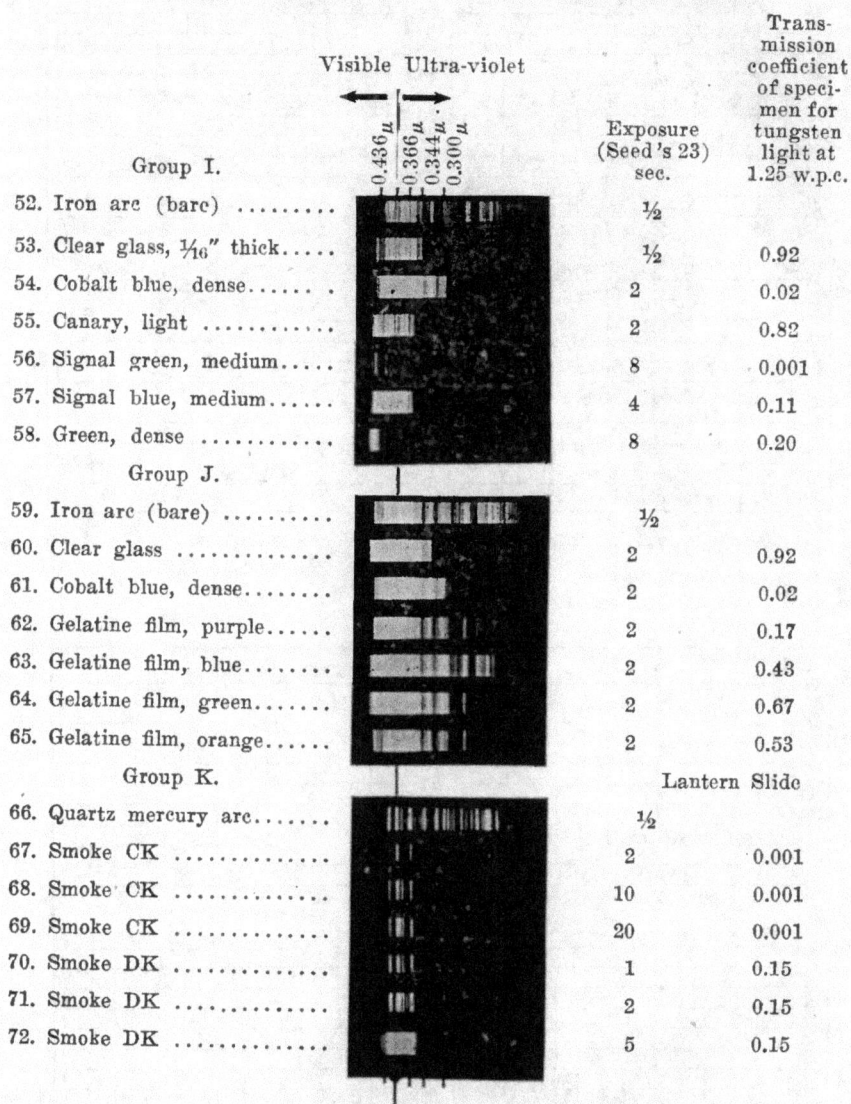

Fig. 22—Spectrograms of the transmission of various glasses in the ultraviolet.
(Courtesy of M. Luckiesh.)

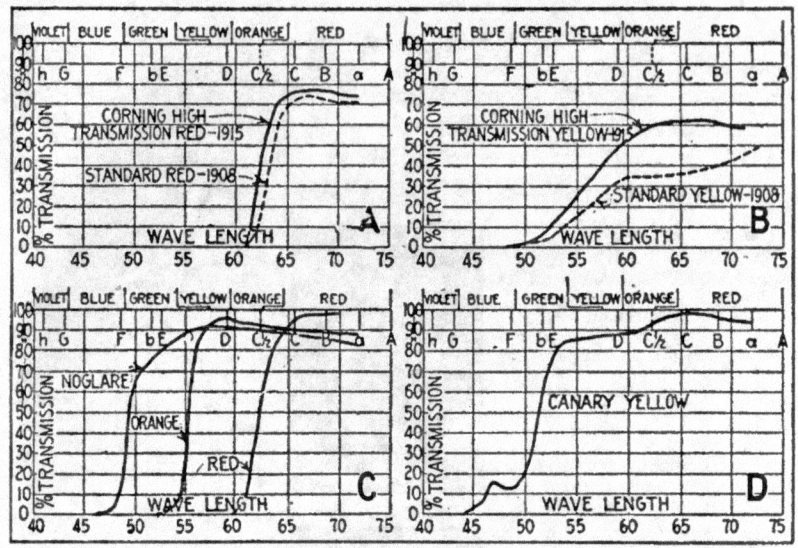

Fig. 23—Characteristic transmission curves for colored glasses. (From Gage: *Trans. Ill. Eng. Soc.*, Vol. XI, 1916.)

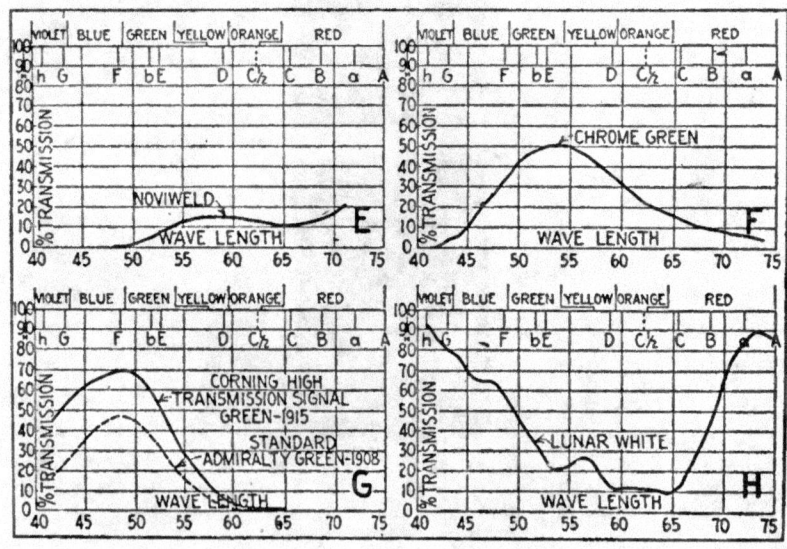

Fig. 24—Characteristic transmission curves for colored glasses. (After Gage: *Trans. Ill. Eng. Soc.*, Vol. XI, 1916.)

parency of clear lead glass remains unchanged to rays as short as 0.35µ. The smoke glasses are representative of many examined. They show little tendency to selectively absorb ultra-violet rays and differ considerably in their characteristics. These glasses cannot conscientiously be recommended with safety for the protection of the eyes against excessive ultra-violet radiation. The amethyst glass absorbs more ultra-violet than clear glass yet is transparent far into the ultra-

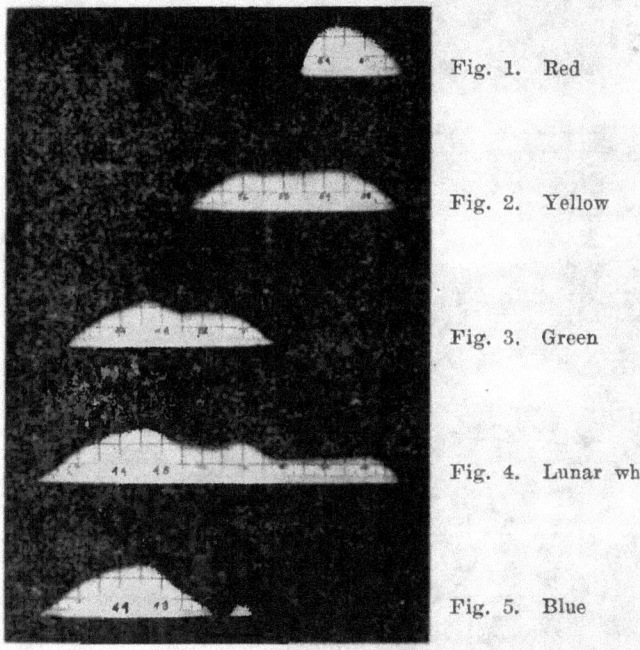

Fig. 1. Red

Fig. 2. Yellow

Fig. 3. Green

Fig. 4. Lunar white

Fig. 5. Blue

Fig. 25—Spectrograms of colored glasses used in railway signals. (After Gage: *Trans. Ill. Eng. Soc.*, Vol. XI, 1916.)

violet region. The amber glasses quite satisfactorily absorb ultra-violet rays but give rise to some objection from the standpoint of color. Several deep-red glasses were examined and all found to be opaque to ultra-violet rays but on account of the strong color should not be recommended. * * * The transmission of Euphos glass decreases considerably in the ultra-violet but shows a tendency to increase in transparency in the region 3200 t. m. This transparency to short wave ultra-violet rays becomes quite marked in less dense specimens.''

Figures 20, 21 and 22 give a number of spectrograms published by Luckiesh illustrating the transparency of various media to rays of different wavelength. The division line between the visible and ultra-violet is set at about 4000 to 3900 t. m. The seconds' exposure and the transmission coefficient for the total visible light from a tungsten lamp are indicated. "In group I and J," writes Luckiesh, "are the

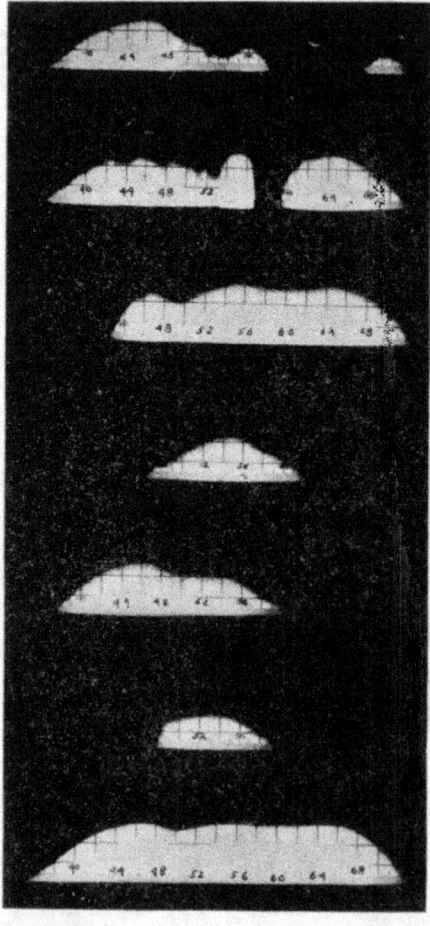

Fig. 1.   Cobalt

Fig. 2.   Didymium

Fig. 3.   Uranium

Fig. 4.   Chrome green

Fig. 5.   Blue green

Fig. 6.   Blue green plus noglare

Fig. 7.   Clear

Fig. 26—Colored glasses exhibiting band spectra (Nos. 1 to 3). Green glasses (Nos. 4 to 6). (After Gage: *Trans. Ill. Eng. Soc.*, 1916.)

spectrograms of various common glasses which are often used in the industries. It is seen that the cobalt-blue glass is more transparent to ultra-violet radiation than is a clear glass of the same thickness. The clear glass used was a lantern-slide cover glass. This difference is best shown in 60 and 61 where the exposures were equal. It is significant to note that the clear glass is approximately 46 times the more transparent to visible light than the cobalt-blue glass.''

Figures 23-26 are taken from an article by Dr. H. P. Gage on ''Colored Glass in Illuminating Engineering,'' (*Trans. Ill. Eng. Soc.,* Vol. XI, 1916). The first two of these diagrams give some characteristic curves of transmission in the visible regions by various colored glasses. Figure 25 gives reproduced spectrograms of the colored glasses commonly used in railway signals. Figure 26 is of interest in that it shows the effects of the addition of certain ingredients to white glass upon the transmissive properties of the product.

The transmission curves of a considerable number of neutral and colored glasses have been determined in the laboratories of the American Optical Company and published in a brochure entitled *"The Ophthalmic Use of Crookes Lenses."* These curves for clear glass, Crookes A and B, a couple of ambers, amethyst, smoke, etc., together with the data giving the approximate absorptions are shown in Figures 27-35, inclusive. It will be noted that the transmission in the visible spectrum for white glass is practically 92 per cent., the reflection from the surfaces amounting to about 8 per cent. The ultra-violet and near violet transmission is shown by shading in two different manners, that lying close to the violet (3900 to 3700 t. m. roughly), and that below 3700 t. m. The transmission for white glass becomes zero at about the 2800 t. m. point. It will also be noticed that the transmission in the ultra-violet is considerably greater than that of any of the other glasses shown. The Wellsworth Crookes A—after the formula of Sir William Crookes—is practically a colorless glass and yet the absorption of the ultra-violet in comparison to white glass is marked, absorbing as it does the ultra-violet completely below 3600 t. m. Another characteristic of both the Crookes A and B shades is the appreciable absorption of a selective character in the yellow region and just above the point generally specified as being the maximum of the sensibility curve of the average eye (5600 t. m.) under fairly high illuminations. The reader will likewise be interested in comparing these curves for Wellsworth Crookes lenses with determinations made at the Bureau of Standards by Gibson and McNicholas (Figure 36). And again, a comparison of Wellsworth Crookes A and B, either in the diagrams accompanying this discussion or in those given by Gibson and McNicholas, show that the limits of absorption in the ultra-violet of the two shades is the same and that the transmission curves are practically one and the same beyond 3700 t. m. Figure 36, by Gibson and McNicholas, gives the transmission curves of white glass, Crookes A Wellsworth, Crookes B Wellsworth side by side upon the same plat

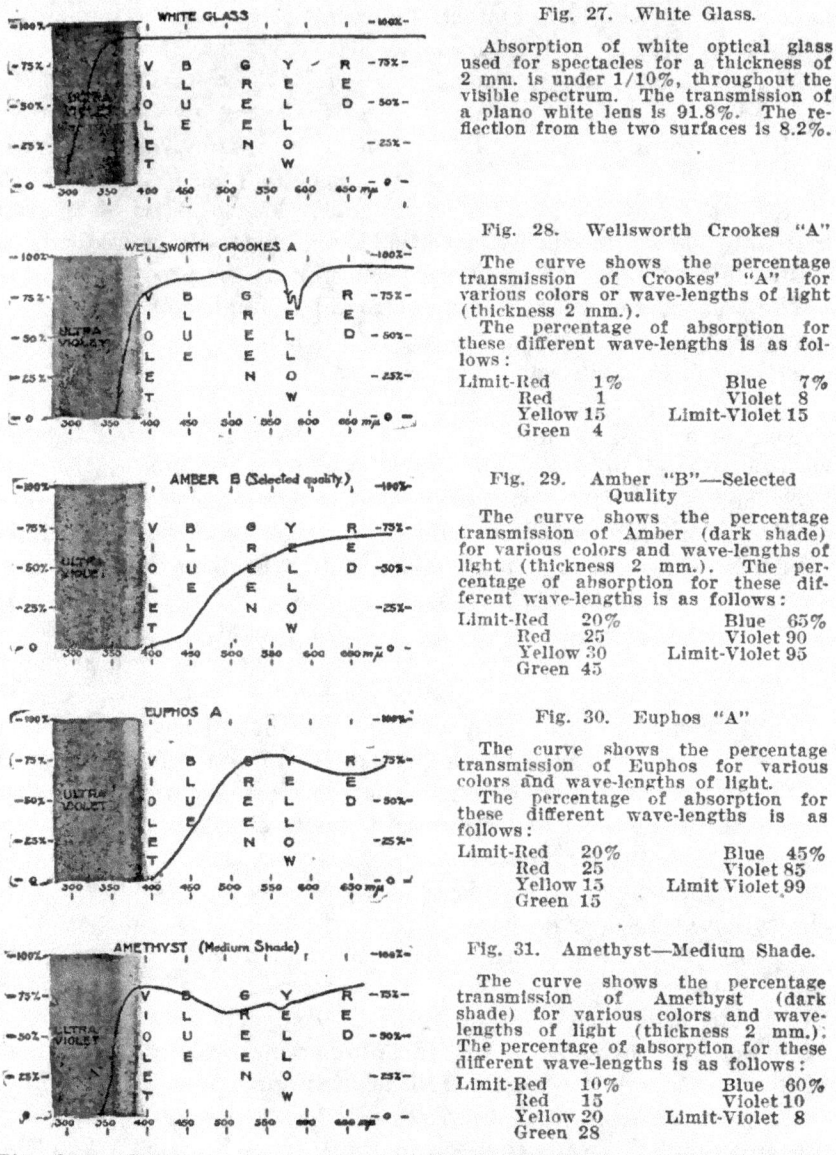

Fig. 27.   White Glass.

Absorption of white optical glass used for spectacles for a thickness of 2 mm. is under 1/10%, throughout the visible spectrum. The transmission of a plano white lens is 91.8%. The reflection from the two surfaces is 8.2%.

Fig. 28.   Wellsworth Crookes "A"

The curve shows the percentage transmission of Crookes' "A" for various colors or wave-lengths of light (thickness 2 mm.).
The percentage of absorption for these different wave-lengths is as follows:

Limit-Red	1%	Blue	7%
Red	1	Violet	8
Yellow	15	Limit-Violet	15
Green	4		

Fig. 29.   Amber "B"—Selected Quality

The curve shows the percentage transmission of Amber (dark shade) for various colors and wave-lengths of light (thickness 2 mm.). The percentage of absorption for these different wave-lengths is as follows:

Limit-Red	20%	Blue	65%
Red	25	Violet	90
Yellow	30	Limit-Violet	95
Green	45		

Fig. 30.   Euphos "A"

The curve shows the percentage transmission of Euphos for various colors and wave-lengths of light.
The percentage of absorption for these different wave-lengths is as follows:

Limit-Red	20%	Blue	45%
Red	25	Violet	85
Yellow	15	Limit Violet	99
Green	15		

Fig. 31.   Amethyst—Medium Shade.

The curve shows the percentage transmission of Amethyst (dark shade) for various colors and wave-lengths of light (thickness 2 mm.). The percentage of absorption for these different wave-lengths is as follows:

Limit-Red	10%	Blue	60%
Red	15	Violet	10
Yellow	20	Limit-Violet	8
Green	28		

Figs. 27-31—Percentage transmission of light by ophthalmic glasses.   (From the *American Optical Co.*)

and this makes comparison easy. Another interesting comparison is that between Crookes B and Smoke B (*vide* Figures 32 and 35). In many respects these glasses are similar in their absorptions but there is one marked and noteworthy difference: the absorption of the ultra-

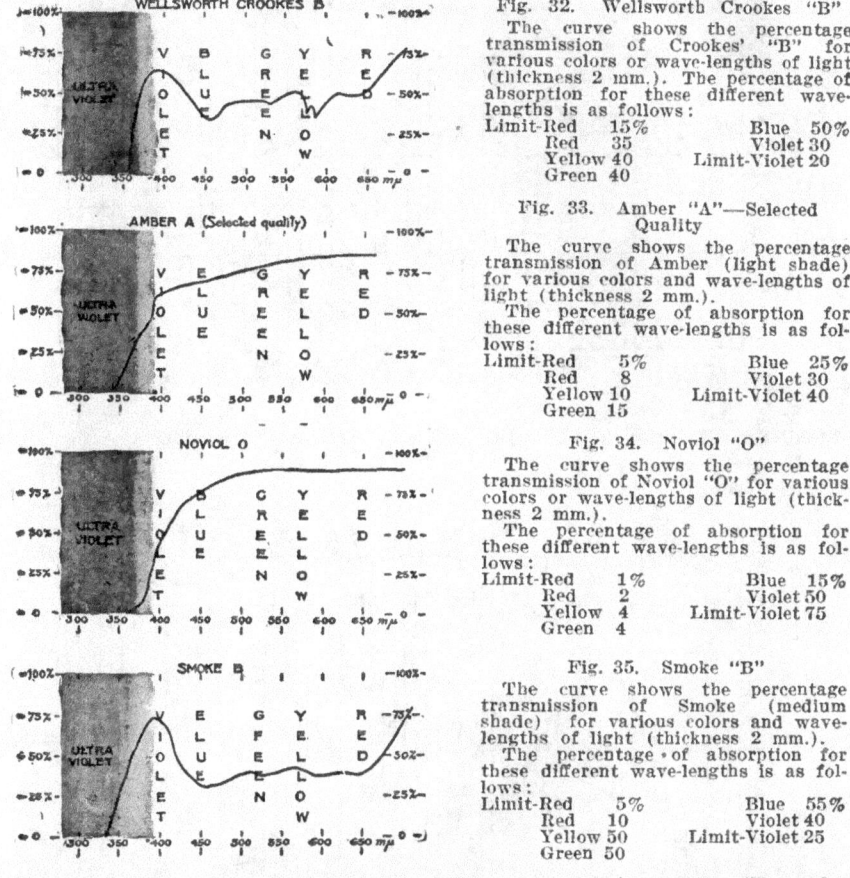

Fig. 32.   Wellsworth Crookes "B"

The curve shows the percentage transmission of Crookes' "B" for various colors or wave-lengths of light (thickness 2 mm.). The percentage of absorption for these different wave-lengths is as follows:

Limit-Red	15%	Blue 50%
Red	35	Violet 30
Yellow	40	Limit-Violet 20
Green	40	

Fig. 33.   Amber "A"—Selected Quality

The curve shows the percentage transmission of Amber (light shade) for various colors and wave-lengths of light (thickness 2 mm.).

The percentage of absorption for these different wave-lengths is as follows:

Limit-Red	5%	Blue 25%
Red	8	Violet 30
Yellow	10	Limit-Violet 40
Green	15	

Fig. 34.   Noviol "O"

The curve shows the percentage transmission of Noviol "O" for various colors or wave-lengths of light (thickness 2 mm.).

The percentage of absorption for these different wave-lengths is as follows:

Limit-Red	1%	Blue 15%
Red	2	Violet 50
Yellow	4	Limit-Violet 75
Green	4	

Fig. 35.   Smoke "B"

The curve shows the percentage transmission of Smoke (medium shade) for various colors and wave-lengths of light (thickness 2 mm.).

The percentage of absorption for these different wave-lengths is as follows:

Limit-Red	5%	Blue 55%
Red	10	Violet 40
Yellow	50	Limit-Violet 25
Green	50	

Figs. 32-35—Percentage transmission of light by ophthalmic glasses. (From the *American Optical Co.*)

violet by Crookes B is complete at slightly under 3600 t. m., while for the Smoke B the absorption is not complete until about 3200 t. m. is reached. There are many reasons for believing that the deleterious or, to say the least, non-desirable effects of the ultra-violet rays under ordinary, workaday conditions lie in the region just below 3600 t. m. Since smoke glass transmits these shorter ultra-violet rays it would appear fairly conclusive that the value of the use of Crookes, Smoke

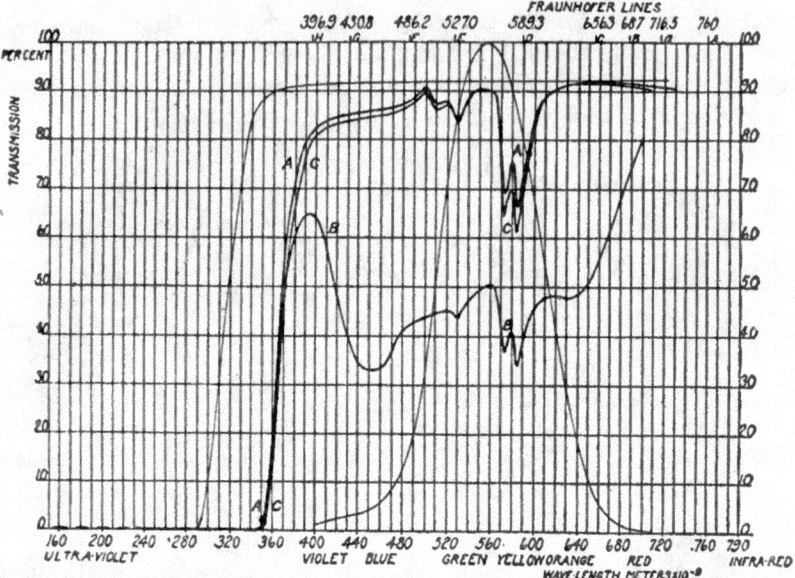

Fig. 36—A="Crookes A, Wellsworth," 2.05 mm; B="Crookes B, Wellsworth," 2.16 mm; A. O. Co. C="91 B," 1.97 mm; Corning. (After Gibson and McNicholas. Permission of Bureau of Standards.)

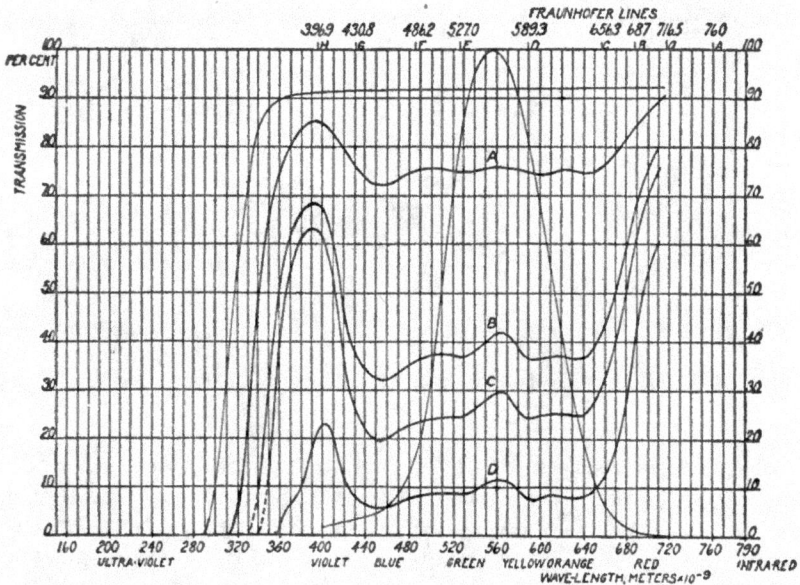

Fig. 37—A="Smoke A," 2.10 mm; B="Smoke B," 2.14 mm; C="Smoke C," 2.13 mm; D="Smoke D," 2.03 mm; A. O. Co. (After Gibson and McNicholas. Permission of Bureau of Standards.)

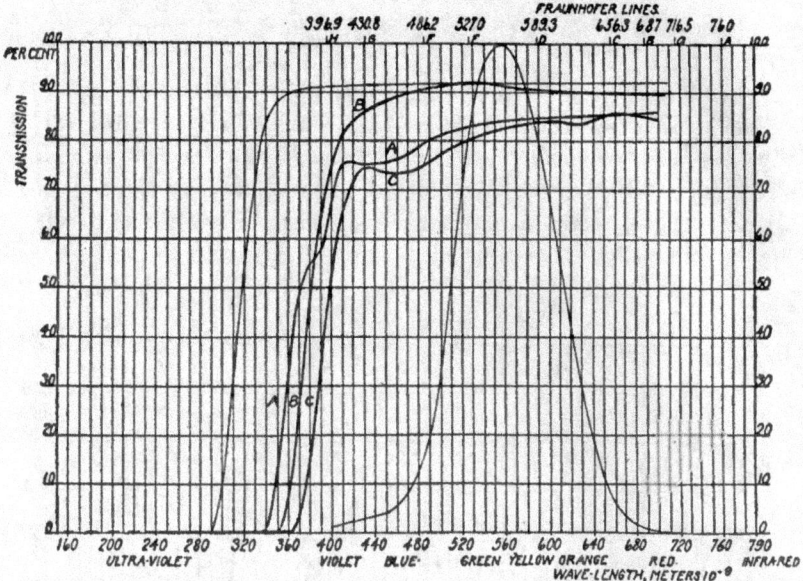

Fig. 38—A="Luxfel," 2.00 mm; B="Lab. No. 57," 1.90 mm; C="Lab. No. 58," 2.02 mm; A. O. Co. (After Gibson and McNicholas. Permission of Bureau of Standards.)

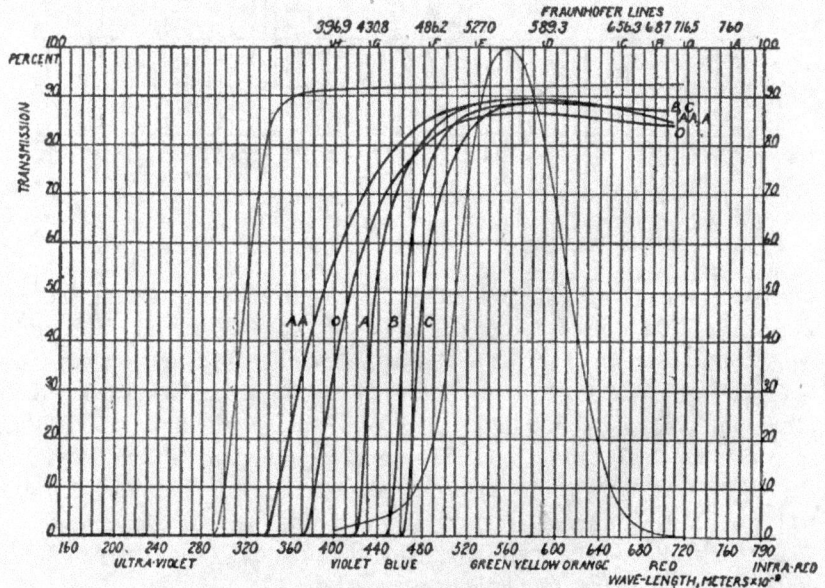

Fig. 39—AA="Noviol AA," 1.88 mm; O="Noviol O," 2.03 mm; A="Noviol A," 2.06 mm; B="Noviol B," 2.10 mm; C="Noviol C," 2.10 mm; A. O. Co. (Permission of Bureau of Standards.)

or similar glasses lies in the reduction of the total quantity of energy entering the eye. A comparison of the data of the curves given in Figures 36 and 37 will emphasize these points of similarity and dissimilarity. Amber A (Fig. 33), Amber B (Fig. 29) and Noviol O (Fig. 34) may properly be grouped in a family for the purposes of discussion. All of these (and other glasses such as Nactic, Luxfel, Oliveye, etc.) possess in general a yellowish or yellowish-green hue. The Noviol O, of all those specified, has the least effect upon the visible

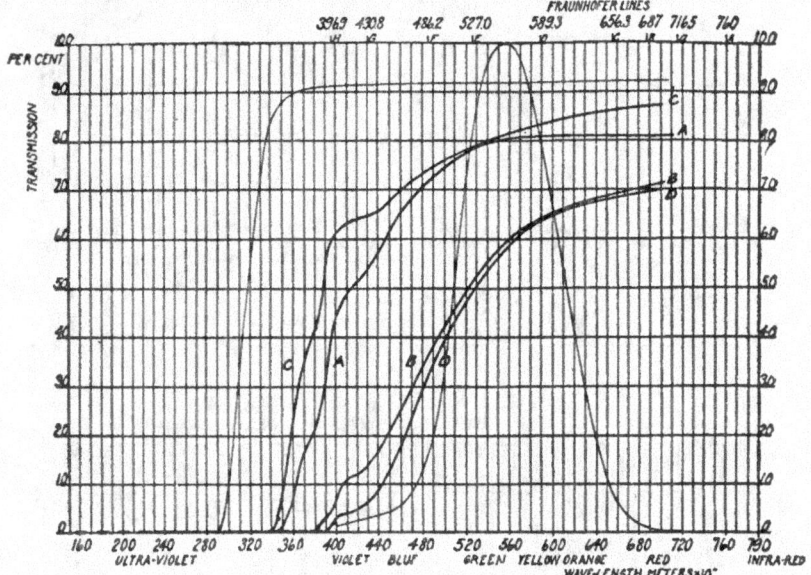

Fig. 40—A="Amber A," 2.13 mm; B="Amber B," 2.13 mm; A. O. Co. C= "Amber, light," 1.93 mm; D="Amber, dark," 1.85 mm; W. & O. (Permission of Bureau of Standards.)

spectrum, since it transmits about 85 per cent. of all wavelengths from the red at 7200 t. m. down to the green-blue at 5000 t. m. The amber A (selected quality, Fig. 33) transmits considerably more ultra-violet than does the Noviol. In turn, however, the amber B absorbs all the ultra-violet beyond 3800-3900 t. m. and has much more marked absorptive effects in the violet, blue and green than either Noviol or the lighter ambers. The transmission curves for Euphos, Fieuzal, Chlorophil, Hallauer, Akopos, Saniweld, and special glasses such as "392 F," "124 J. A.," etc., and the Pfund gold film between plates of Crookes glass are given in Figures 41 to 44.

Reference has already been made to the work of Gibson and Mc-Nicholas on the Ultra-violet and Visible Transmission of Eye-Pro-

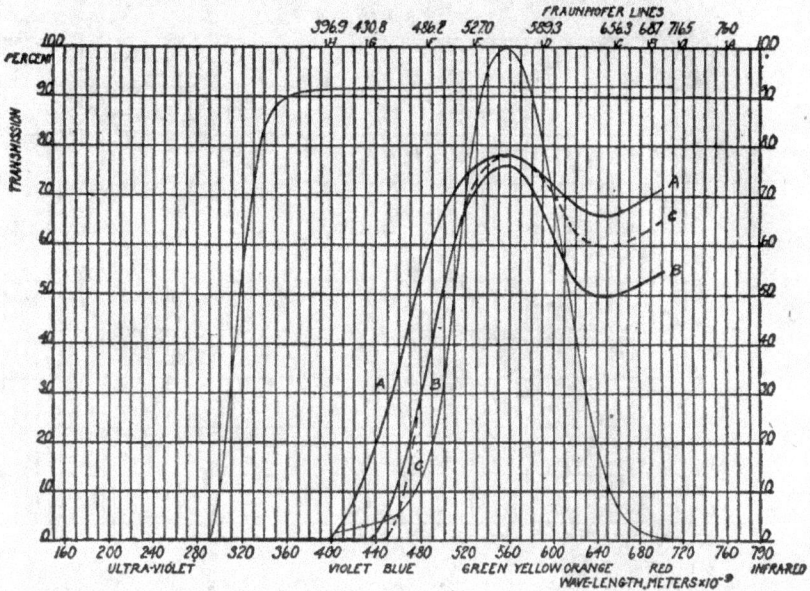

Fig. 41—A="Euphos," 1.95 mm; B="Lab. No. 61," 2.13 mm; A. O. Co. C=glass labeled "Fieuzal," bought in store. (Permission of Bureau of Standards.)

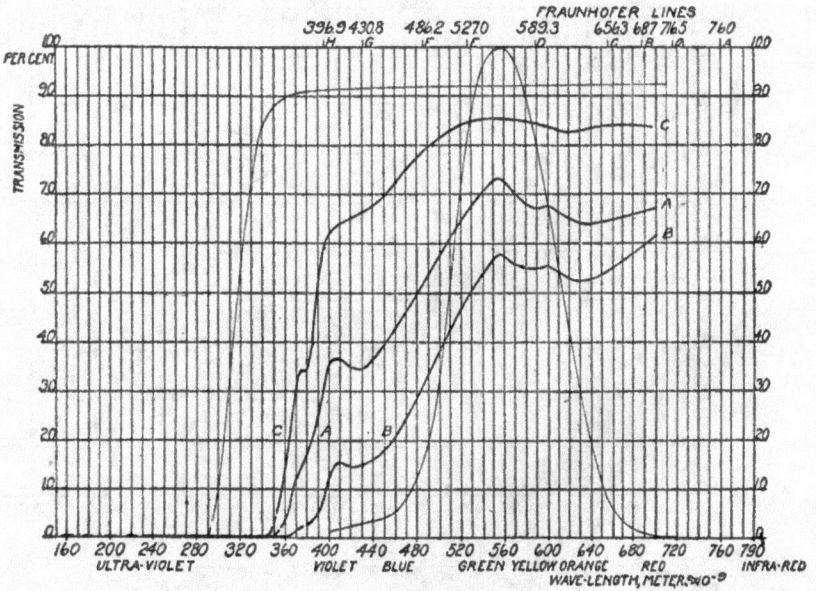

Fig. 42—A="Fieuzal A," 2.13 mm; B="Fieuzal B," 2.13 mm; A. O. Co. C="Fieuzal," 1.98 mm, W. & O. (Permission of Bureau of Standards.)

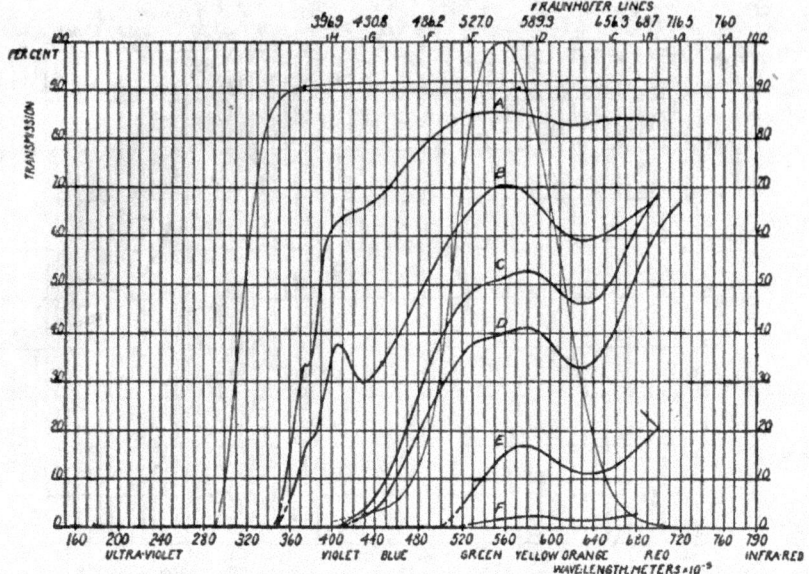

Fig. 43—A="Fieuzal," 1.98 mm; B="Chlorophile," 1.98 mm; C="Hal-lauer," 1.90 mm; W. & O. D="Akopos," 2.17 mm; E="Saniweld, light," 1.82 mm; F="Saniweld, dark," 2.12 mm (see Fig. 21); King. (Permission of Bureau of Standards.)

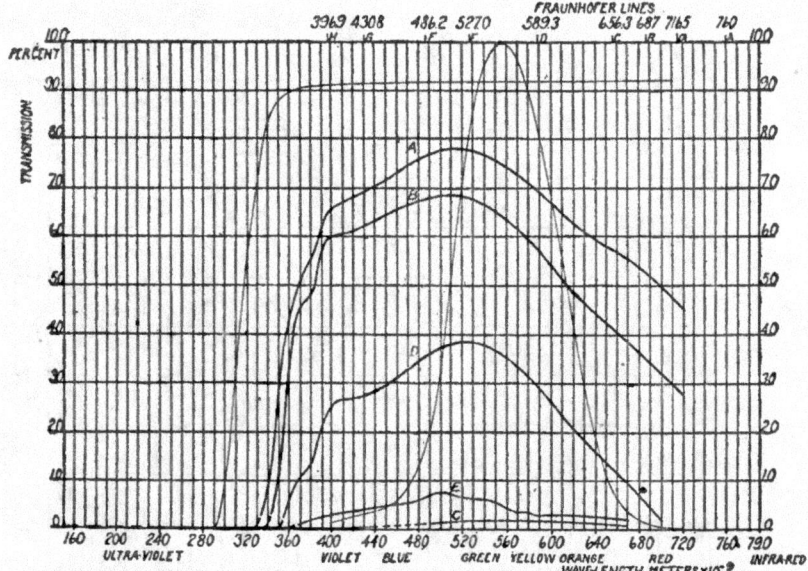

Fig. 44—A="392 F," 1.90 mm; B="124 JA," 2.02 mm; C="124 IP," 2.00 mm (see Fig. 21); Corning. D="Lab. No. 59," 2.13 mm; E="Pfund," gold film between plates of "Crookes" glass, total 2.89 mm; A. O. Co. (Permission of Bureau of Standards.)

tective Glasses. This appeared as one of the *Technologic Papers of the Bureau of Standards* in 1919, (No. 119). Without doubt it is the most exhaustive study of the subject yet made. For details of the experimental procedure and for the methods they devised of computing the transmission for different thicknesses of glass, the reader is referred to the original paper. Figures 36 to 47 inclusive are taken from the paper by Gibson and McNicholas and are self-explanatory. In commenting upon the results of their investigations these men

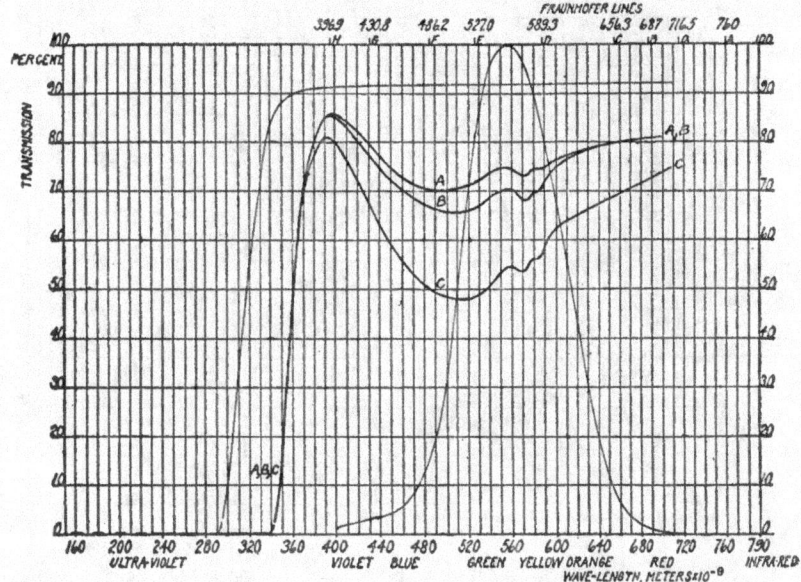

Fig. 45—A="Amethyst A," 2.08 mm; B="Amethyst B," 2.05 mm; C="Amethyst C," 2.04 mm; A. O. Co. (Permission of Bureau of Standards.)

write: "Of the specimens studied, the five kinds which are most efficient as protection against the ultra-violet, while being at the same time nearly colorless in the thicknesses examined, are 'Crookes A,' Corning '91 B,' A. O. Co. Lab. 'No. 57,' A. O. Co. Lab. 'No. 58,' and 'Noviol O.' Of these, 'Noviol O' and A. O. Co. 'Lab. No. 58' are the best, but are not so truly colorless as the other three. Of the slightly colored glasses, by far the best seem to be 'Noviol A' and 'Noviol A_1,' as they absorb completely below 410 mμ. while transmitting about 87 per cent. of the incident light. It is not thought that the slight color would be at all objectionable for ordinary use. A combination of 'Noviol A' and Corning '124JA' is very efficient for eye protection, as it absorbs all the ultra-violet and most of the infra-red, and still has high visible transmission. The color is a very light green and the colors of objects viewed through it are distorted practically none at all. A gold film on 'Noviol A' glass would also be very efficient, though transmitting less of the visible than the combination just mentioned. The yellow and yellow-green glasses of a deeper shade are usually good protection against the ultra-violet. The green and blue-green glasses of Fig. 44 are used primarily to protect the eye from the infra-red. The 'Pfund' specimen is a gold

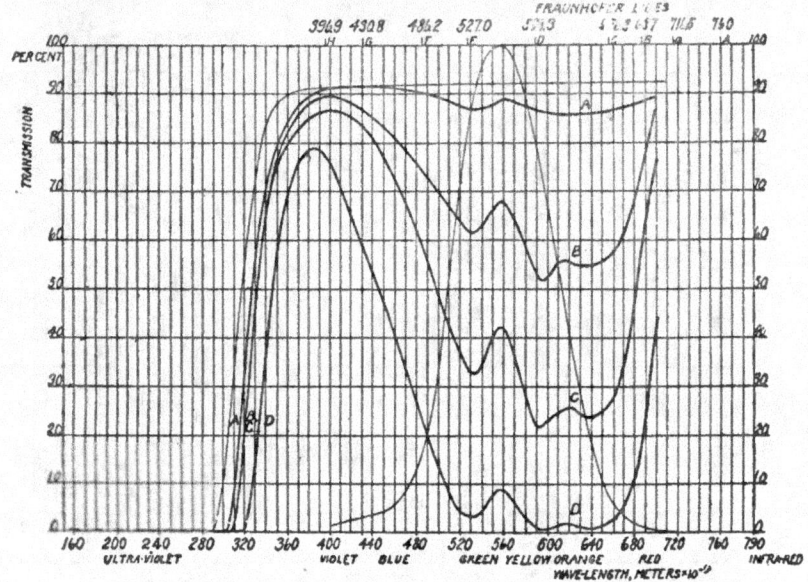

Fig. 46—A="Blue A," 2.10 mm; B="Blue B," 2.04 mm; C="Blue C," 2.05 mm; D="Blue D," 2.11 mm; A. O. Co. (Permission of Bureau of Standards.)

film between two pieces of what seems to be 'Crookes' glass. 'Smoke,' amethyst and blue or purple glasses are liable to be little better than clear glass as a protection against the ultra-violet. Of the welding glasses, yellow seems to be the safest, as the green or neutral shades are liable to have transmission bands centering near 395 $m\mu$, which may extend to a considerable distance into the ultra-violet.

Gibson and McNicholas also investigated the transmissions of a few glasses which may be classed as welding glasses. These glasses are for use under high powered arcs and chiefly in industries in which welding enters. Figures 48, 49 and 50 give the graphical results in the ultra-violet and visible of several kinds of special welding glasses.

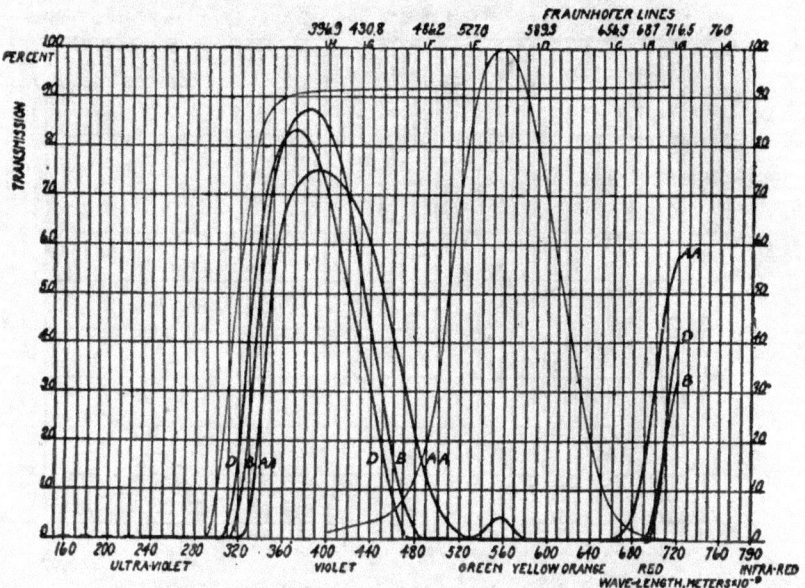

Fig. 47—AA="Cobalt Blue AA," 2.75 mm; B="Cobalt Blue B," 1.85 mm; D="Cobalt Blue D," 1.86 mm; "Cobalt Blue A," 3.20 mm nearly same as curve AA; "Cobalt Blue C," 1.46 mm nearly same as curve B; "Chromatic Test," 2.36 mm similar to curve B, but slightly lower in value; A. O. Co. (Permission of Bureau of Standards.)

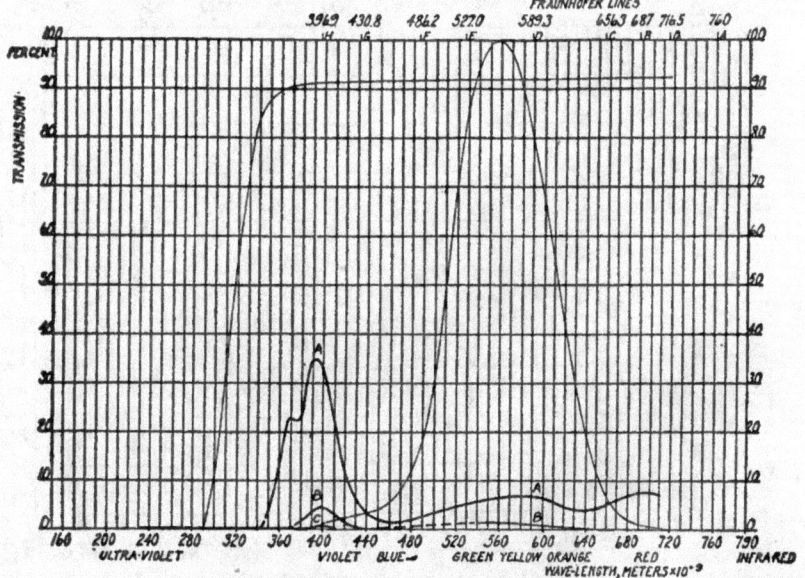

Fig. 48—A="Welders Smoke Dark," 1.42 mm; B="Special Welders Light," 1.68 mm (see Fig. 21); C="Special Welders Dark," 2.54 mm (see Fig. 21); A. O. Co. (Permission of Bureau of Standards.)

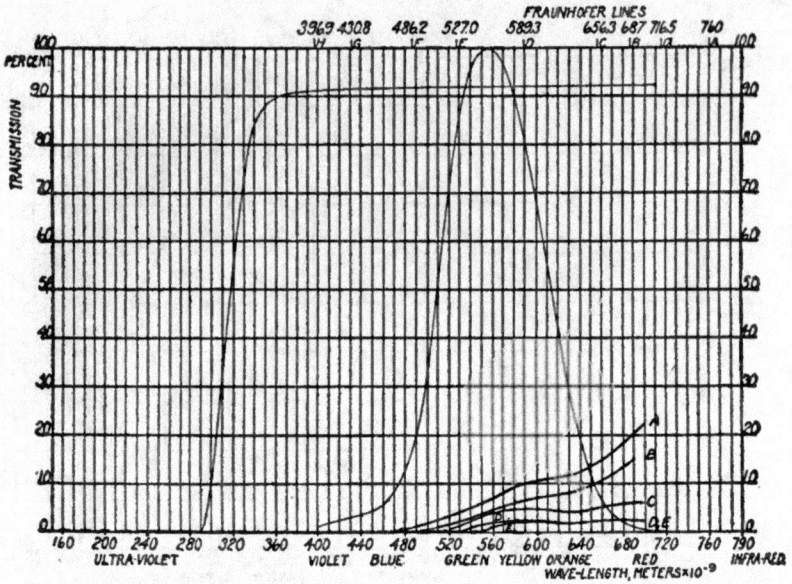

Fig. 49—A="Welding glass 6," 1.97 mm; B="Welding glass 2," 2.32 mm; C. E. S. Co. C="Noviweld 4," 1.89 mm; D="Noviweld 5," 2.16 mm; E="Noviweld 6," 2.20 mm; "Noviweld 7," 1.90 mm 1.02% at 578; "Noviweld 8," 2.01 mm 0.517% at 578; A. O. Co. (Permission of Bureau of Standards.)

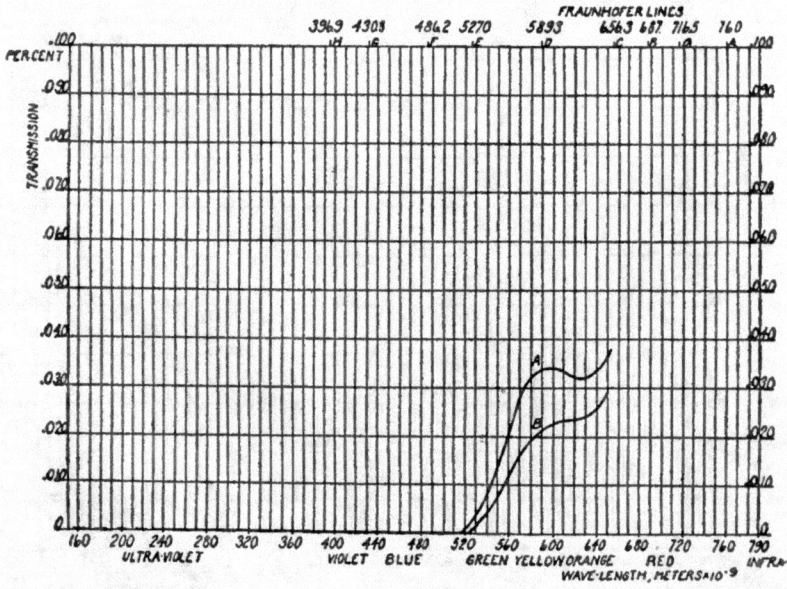

Fig. 50—Note scale of ordinates. A="Special Noviweld No. 8," 1.77 mm; Hardy. B="391 DD," 1.90 mm; Corning. (Permission of Bureau of Standards.)

The extremely small percentage (about 0.03 per cent.) of transmission in the yellow-green region for the special Noviwelds shown in Figure 50 is worthy of notice.

One criticism which it seems to the writer may be passed upon the work of Gibson and McNicholas is that, in the majority of their tests, transmission measures in the ultra-violet were not carried further than 2 per cent. With low incident energy or brief times of exposure this small amount of transmission in the extreme ultra-violet obtained with glasses, etc., might be neglected, but this low percentage transmission may still be extremely important under conditions imposed by welding operations and so forth.

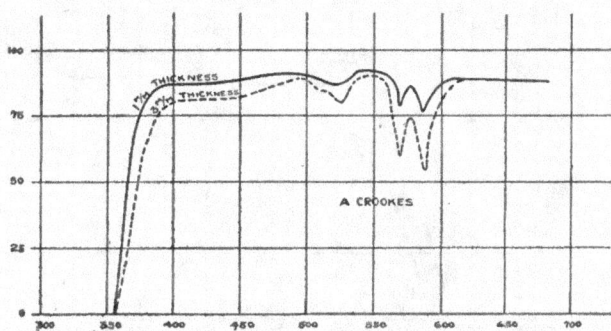

Fig. 51—Effect of thickness on transmission of Wellsworth Crookes glass.
(Permission of the American Optical Company.)

In all of these later experiments upon the transmission of glass due care has been taken to have all the results reduced to a uniform basis; i. e., the element of variation of thickness of sample of glass has been eliminated. It is of interest, therefore, to see what effects thickness has upon the transmission of glass having thicknesses found in ophthalmic lenses. An article by Sheard on the Effect of the Thickness of Glass upon the Transmission of Various Parts of the Spectrum (*Wellsworth,* page 140, 1919) gives the results upon Crookes A, Noviol O, Smoke B and Fieuzal A. Every lens other than a plano has a thickness varying from the center to the edge. In the case of convex lenses, the thickness increases toward the position of the optical center: in concave lenses just the reverse is true. Hence, a high-powered plus lens may be several millimeters in thickness at the optical center, while a concave lens may be almost as thin as a sheet of paper at the center point. It necessarily follows, therefore, that the tint of a lens cannot be preserved uniformly over the surface of lenses having appreciable power. An amber lens, for example, becomes lighter and the color fades out at the thinner portions of the finished lens, although the original glass block from which the lens was manufactured was of uniform color or tint throughout. As a result therefore, the color and the percentage of transmitted light of various wavelengths vary. There is, possibly, one method which would annul such effects and that is the scheme of applying the tint, somewhat after the manner of a coat of paint, to the surfaces of the lenses after they are finished. But no satisfactory device has yet been discovered which will give the effect and the durability needed. And it is to be seriously doubted whether this process of coating lenses with desired colors, even if discovered, would obviate the changes

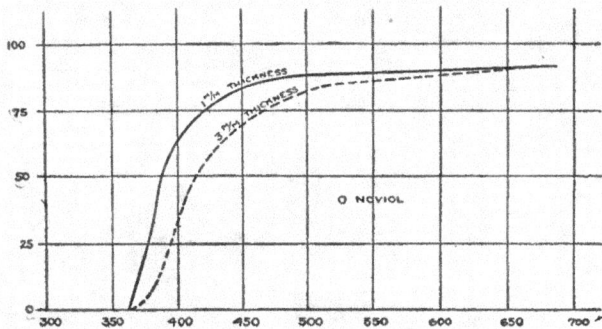

Fig. 52—Effect of thickness on the transmission of Noviol O. (From the American Optical Company.)

in shade or tint, since the same fundamental problem of absorption of glass as dependent upon its thickness would again enter.

The four sets of curves given in Figures 51-54 were determined in the Research Laboratories of the American Optical Company. They are given as typical curves illustrative of the effects of glass upon the transmission of spectral energy from the red down through and including the ultra-violet. In each of these diagrams the upper curves represent the transmission through one centimeter of glass and the lower curves give the results when the thickness of glass is of the order of three millimeters.

The curves show that a change of thickness from 1mm. to 3 mms. has but little effect upon the transmission of the visual and the ultra-violet radiations in the case of Crookes A and Noviol O. The greatest effect of thickness in Crookes A lies in the yellow-green region (5500 to 6000 Angstroms) where the absorption in the two characteristic bands is increased almost 25 per cent. The average effect of tripling the thickness is about 10 per cent. in the green to ultra-violet regions inclusive. Hence, the most marked effect of change of thickness in Crookes A lies in the increased absorption in the yellow-green region; i. e., the region of maximum visibility of the human eye. Furthermore, the change in tint due to varying thicknesses of glass in the finished product is practically negligible with Crookes A and Noviol O.

Noviol O shows considerable increase (about 15 per cent.) in the absorption of the short wavelengths as the thickness is changed from 1 mm. to 3mms. Thickness of glass has but little effect upon the transmission of the yellowish-green, yellow, orange and red radiations in the case of Noviol O. Hence this glass, by increased thickness, cuts down the percentage transmission of the green, blue, violet

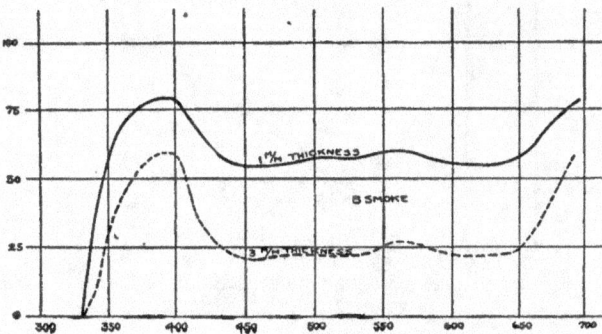

Fig. 53—Effect of thickness on the transmission of B Smoke. (From the American Optical Company.)

and ultra-violet without appreciably affecting the transmission of energy lying in the region of maximum visibility.

The data on Smoke B and Fieuzal A, as plotted in Figures 53 and 54, speak for themselves. In the case of Smoke B, in the region from 6500 to 4500 Angstroms approximately, a change of thickness from 1 mm. to 3 mms. changes the transmission from 50-60 per cent. to 20-25 per cent. The effect of thickness is less marked in the ultra-violet regions. It follows, therefore, that thickness has a marked effect upon the transmission throughout the whole of the visual spectral region.

Fieuzal A shows that the effects of thickness are pronounced, especially in the region from 5500 to 4000 Angstroms. Here again the percentage transmission is cut in half by a change of thickness from 1 mm. to 3 mms. of glass.

The general conclusion which may be drawn from these curves is: Appreciable changes in thickness—as judged from the standpoint of ophthalmic lens manufacture—may occur in lenses and prisms made from Crookes A and Noviol O without any marked change in the percentages of various kinds of radiant energy transmitted by these glasses. Hence, they are nearly as effective in their transmissive and absorptive powers when made up in lenses having thicknesses up to 1 mm. as when these thicknesses are of the order of 3 mms. As a result, we are led to conclude that concave or convex lenses of high powers will have, when made up in Crookes A and Noviol O, practically the same effect upon the character of the luminous energy ultimately reaching the retina. Therefore, thicknesses commonly used in ophthalmic lens manufacture will not cause any noticeable

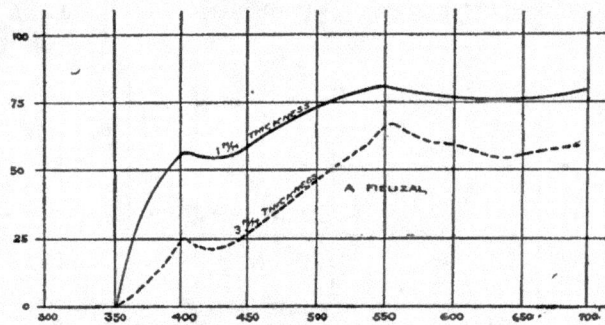

Fig. 54—Effect of thickness on the transmission of A Fieuzal. (Permission of the American Optical Company.)

variation in the tint or affect the transmission in appreciable amounts if either Crookes A or Noviol O is used.

The Infra-Red.

The fact that glassblowers are subject to a special form of cataract has raised the question as to whether or not this action is due to radiant energy and if so, whether this action is of an abiotic or thermic nature or whether it is caused by ultra-violet or infra-red radiations. Meyhofer in 1886 examined over five hundred glassmakers and found about 12 per cent. affected with cataract. The cataract almost always appears first in the left eye which is the more exposed to the energy from the molten glass. When appearing in the right eye first Stein showed that the glassblower had the habit of turning that side of the face toward the oven. The length of time necessary for development of the cataract is not exactly known but evidently comprises several years. The workmen also develop a peculiar rusty-brown spot on each cheek, generally more noticeable on the left. Hirschberg states that for over one hundred years it has been known that individuals exposed to intense heat and light are especially liable to cataract. In the case of the glassblowers, the cataract begins as a rosette-like or diffuse opacity in the cortex at the posterior pole of the lens. Later, striæ similar to those of senile cataract may appear. The great frequency, therefore, with which glassblower's cataract occurs, its relatively uniform character, and the fact that it occurs first in the more exposed eye argues for the statement that the cataract is due to radiant energy on the eye itself. The further questions as to whether the cataract is due to the direct action of the light upon the lens or upon the eye as a whole, and whether it is due to abiotic or thermic action are not so easily answered. Cramer, Stein and others believe that cataract is due to the chemical action of the ultra-violet; Vogt regards the infra-red as chiefly responsible.

The character of the radiation from molten glass is well known. It is that of an incandescent body of about 1200° to 1400° C. Crookes (*Trans. Royal Soc. Lon.*, 1914) says: "As far as one can judge the temperature at the melting end is about 1500° C. and at the working end decidedly less—say 1200° C." It is certain that the spectrum of a non-gaseous body at this temperature does not include any of the so-called abiotic radiation since the extreme limit of the spectrum of molten glasses as found by investigators is 3200 t. m., and estimates range from that up to 3350 t. m. Crookes (*l. c.*) made six exposures, as reported in his paper, with different times of exposure and found that the spectrum extended to 4520 t. m. after twenty minutes' exposure and from that time of exposure on, the limit of spectrum in the ultra-violet increased to 3345 t. m. after one hundred and eighty minutes' exposure. Certain it is that an exposure of three hours does not permit of the presence of much ultra-violet radiation. Also, without doubt, abiotic action cannot be traced beyond about 3100 t. m. Furthermore, the radiation of a body at such temperatures is relatively weak all through the ultra-violet (*vide* Figure 4). The maximum, according to Planck's and Wein's laws, for a body at 1300° C. lies far in the infra-red, while the energy in the whole visible and ultra-violet part of the spectrum is less than one per cent. of the total. Hence, to ascribe injurious effects to the visible or ultra-violet radiations without the elimination of the 99 per cent. of infra-red radiation would be, on its face at least, to lose all sense of the possible correlation of cause and effect.

These possible thermal effects on the eyes and this abundance of infra-red radiation are of significance in those who engage in such vocations as glassblowing and industries in which welding under powerful arcs is common. We are not desirous at this point of entering into an account of the various arguments and experiments for and against the view that the infra-red radiations produce deleterious actions upon the eye: we shall simply give the results of various investigators as to the absorption by various glasses in the infra-red.

Sir William Crookes appears to have been the first to systematically engage in the development of glasses highly absorptive in the infra-red. His experimentation was carried on in connection with the Glass Workers Cataract Committee of the Royal Society, and consisted in the finding of the effects upon the ultra-violet, the visible and the infra-red of the addition of small quantities of metals such as cerium, chromium, cobalt, copper, iron, lead, manganese, uranium, neodymium, and so forth, to the raw soda flux. He developed a glass No. 246, consisting of 90 per cent. raw soda flux, 10 per cent. ferrous oxalate (a small quantity of red tartar and powdered wood charcoal was added to prevent oxidation) of a sage-green in color which cut off ultra-violet down to 3800 t. m., gave a heat absorption of 98 per cent. and transmitted 27.6 per cent. of the incident light. Another glass, No. 217, prepared from fused soda flux 96.8 per cent., ferroso-ferric oxide 2.85 per cent. and carbon 0.35 per cent., was found to cut off the ultra-violet below 3550 t. m., to cut off 96 per cent. of the heat radiation and to transmit 40 per cent. of the light.

In 1917 W. W. Coblentz and W. B. Emerson of the Bureau of Standards issued a paper on Glasses for Protecting the Eyes from Injurious Radiations (*Technologic Papers, Bureau of Standards*, No. 93). This paper deals largely with the protective properties of glasses which shield the eye from infra-red rays. In order to discuss their results we shall follow their sub-divisions of subject matter according to the color of the glasses. The figures and diagrams accompanying this discussion are from the paper of Coblentz and Emerson.

Curve A (Figure 55) shows the transmission of energy by the human eye. From this transmission curve it will be noticed that radiations of wavelength greater than 1.4 μ cannot reach the retina. In fact, because of the presence of water, which is very opaque to infra-red rays, but little radiation of wavelengths greater than 1.5 μ passes through the cornea. The cornea is about 0.6 mm. thickness. From this it will be noted that about 97 per cent. of the energy radiated from a furnace at 1000° to 1200° C. (*vide* Curve B, Figure 57) is absorbed in the outer portion of the eye.

Yellow-colored glasses. Curve B, Figure 55, gives the transmission of a Noviol glass, curve C that of an orange and curve D that of a canary glass all of 2 mm. thickness. Curve E is that of a colorless (or white) glass. The obstruction of these yellow glasses is but little greater than that caused by an equal thickness of colorless glass.

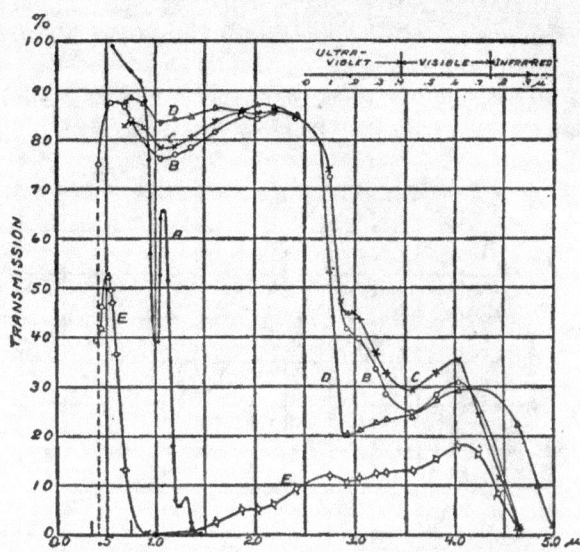

Fig. 55—A, human eye, Corning noviol glasses; B, yellow (thickness, t=2.05 mm); C, orange (t=2.03 mm); D, canary (t=1.85 mm); E, Corning G 124 JA, blue-green (t=1.5 mm). (After Coblentz and Emerson. Permission of Bureau of Standards.)

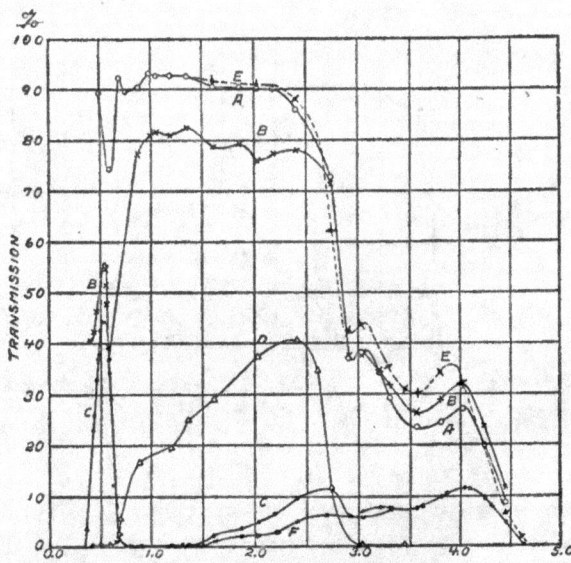

Fig. 56—Crookes's glasses; A, light (t=1.96 mm); B, dark (t=2.00 mm); C, ferrous No. 30, sage-green (t=1.98 mm). D, Schott's black glass (t=3.6 mm). E, white crown glass (t=2.18 mm). F, blue-green glass (A. O. C. Lab. No. 59; t=1.93 mm). (A and B are Crookes's neutral-tint glass.) (After Coblentz and Emerson. Permission of Bureau of Standards.)

The amount of infra-red transmitted by such a glass as Noviol is about 55 per cent. of the total radiation from a furnace heated to 1000° to 1100° C.

Crookes's glasses. Figure 56 gives the transmission curves for a family of Crookes glasses. The lighter or neutral shades absorb but

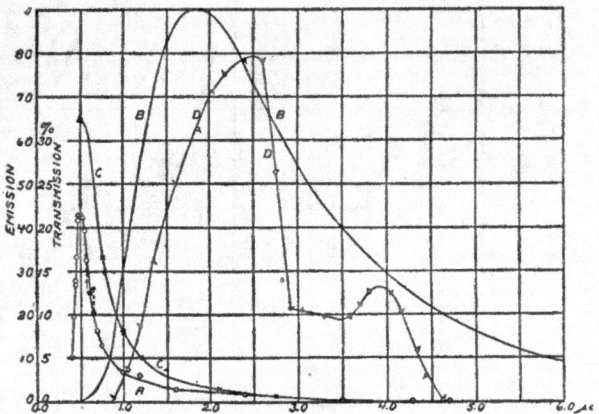

Fig. 57—A, C, gold glass; B, emission of black body (1050° C); D, electric smoke (red) [ordinates=emission scale] (t=2.52 mm). (Permission of Bureau of Standards.)

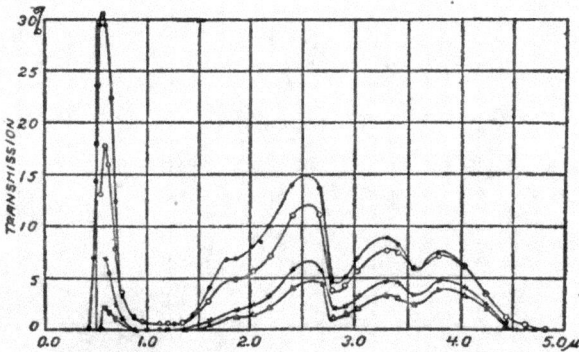

Fig. 58—Corning "New Noviweld" glasses: Top curve=shade 30 per cent; second curve from top=shade 3; next lower curve=shade 4½; bottom curve= shades 6 and 7. Thickness of glasses, 2.2 mm. (Permission of Bureau of Standards.)

little more than white crown glass. Curve C gives the transmission of Crookes sage-green glass (marked Ferrous No. 30). The transmission in the green is about 45 per cent., while in the infra-red the maximum transmission is about 11 per cent. and this for only a narrow spectral region. Curve F is from a blue-green glass (marked

Lab. No. 59 from the American Optical Co.) which transmits about 43 per cent. in the visible. In the infra-red it is more opaque than the sage-green just described.

Pfund gold-plated glass. Metals are extremely opaque to infra-red radiations. In the visible spectrum gold has a region of low reflectivity and great transparency in the region of 0.5 μ (green). This property naturally suggests itself as an excellent method of eliminating all the infra-red by covering white spectacle glass with a thin layer of gold. "The high reflecting power (metallic reflection of 60 to 80 per cent. as compared with the vitreous reflection of about 4 per cent. from glass) makes it desirable to mount these gold-plated glasses in a hood (goggles) which prevents reflection of light from the rear surface of the film into the eye. Curves A and C of Figure 57 show that the gold-plated glass is an extremely effective means of shielding the eye from the infra-red rays. At 1.5 μ the transmission is only about 2 per cent., while beyond 2 μ the transmission is less than 1 per cent. This Pfund glass obstructs 99 per cent. of the infra-red rays emitted by a furnace heated to 1050° C. The Pfund gold-plated glass, made by the American Optical Co., is put out as a gold film deposited upon Crookes A."

Blue-green glasses. Curve E of Figure 55 shows the transmission of a bluish-green glass (Corning G 124 JA) which has fifty per cent. transmission in the green and a very low transmission in the infra-red. This sample transmits only 6 per cent. of the infra-red radiation from a furnace at 1050° C.

Greenish-brown glasses. These glasses protect from the ultra-violet and to some extent from the infra-red rays. The maximum transmission in the visible is about 27 per cent. The coloring matter is effective in its absorption at 1 μ but beyond 3 μ the transmission is about as high as in uncolored glass.

Black glasses. Curve D in Figure 56 gives the transmission of a sample of Schott's black glass: the transmission in the visible spectrum is quite uniform and amounts to about 0.5 per cent. The sample used in Figure 56 transmitted little beyond 3 μ although a lighter colored shade was transparent to 5 μ. This sample transmits about 18 per cent. of the infra-red radiation emitted by a black body heated at 1050° C.

Noviweld glasses. As illustrated in Figure 58, the infra-red transmission of modern noviweld glasses is practically suppressed. The darkest shades transmit only about 1 per cent. of the infra-red radiation emitted from a furnace heated to about 1000° C. The transmission in a rather selective region with a maximum at about 0.5 μ (yellowish-green region) is rather marked.

The transmission curves in the visible and infra-red regions for the French Fieuzal and the German Hallauer glasses are shown in Figure 59.

It will be noted that glasses which absorb highly in the infra-red have either a low transmission throughout the visible spectrum or have the transmission band shifted into the green or blue.

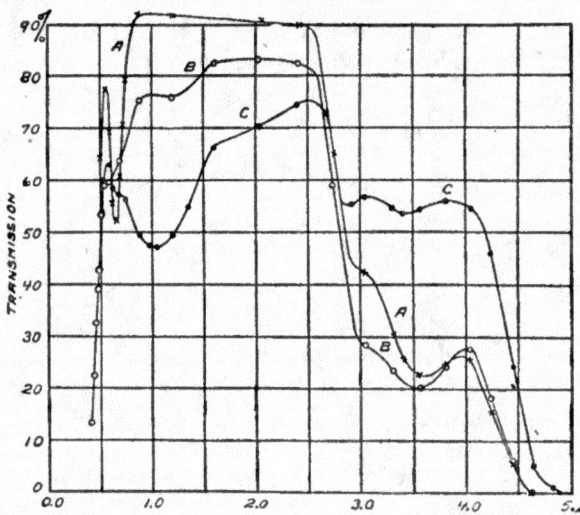

Fig. 59—A, Lab. No. 61, A. O. C. (t=2.09 mm); B, Fieuzal glass, shade B (t=2.04 mm); C, Hallauer glass (t=1.41 mm). (Permission Bureau of Standards.)

Coblentz and Emerson say, by way of conclusion, that "For shielding the eye from infra-red rays deep-black, yellowish-green, sage-green, gold-plated and bluish-green glasses are the most serviceable. For working near furnaces of molten iron or glass, if considerable light is needed a light bluish-green or sage-green glass is efficient in abstracting the infra-red rays. For working molten quartz, operating oxyacetylene or electric welding apparatus, search-lights, or other intense sources of light, it is important to wear the darkest glasses one can use, whether black, green (including gold-plated glasses) or yellowish-green, in order to obstruct not only the infra-red but also the visible and the ultra-violet rays."

Figure 60 gives a good comparative set of curves for the transmission of the eye media, yellow glass, sage-green, neutral tint, gold-plate, greenish-brown, black and blue-green glasses and the emission curves of a black body at 1050° C.

A detailed examination of the infra-red transmission of a considerable number of glasses on the market and used for spectacle lenses was made in 1917 by A. W. Smith and C. Sheard. The results of their investigations are published in the *Journal of the Optical Society of America.* (Vol. II-III, Jan. 1919). The Hilger infra-red

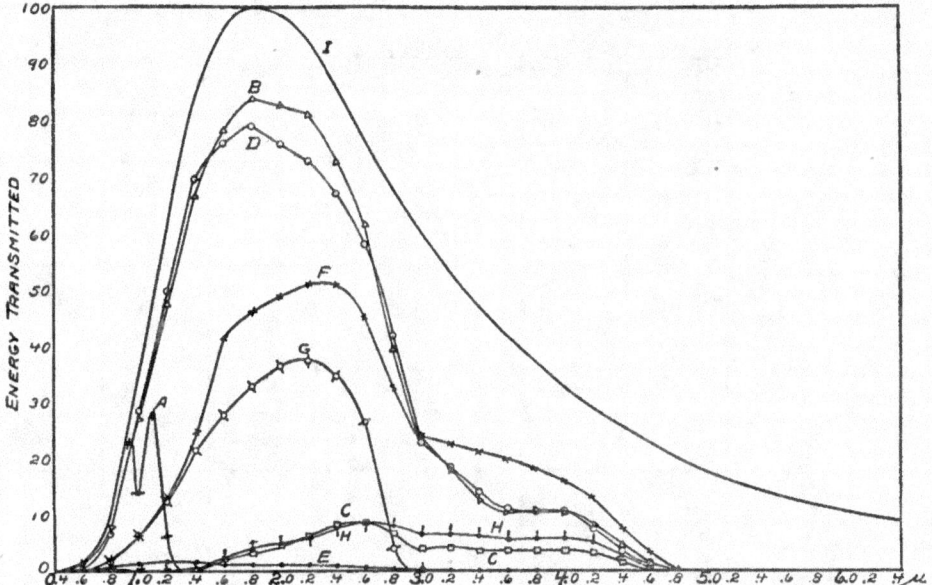

Fig. 60—A, eye media (Fig. 1, A); B, yellow glass (Fig. 1, B); C, sage green (Fig. 2, C); D, neutral tint (Fig. 2, B); E, gold plate (Fig. 3, A); F, greenish-brown (Fig. 4, A); G, black glass (Fig. 2, D); H, blue-green (Fig. 1, E); I, black body (1050° C). (After Coblentz and Emerson. Permission of Bureau of Standards.)

spectrometer was used for these investigations. The width of slits used in these experiments was such as to give a range of spectrum at the thermopile of between 0.1 and 0.26 μ. A Nernst glower served as a radiation source. Two shutters were mounted in front of the spectrometer slit; one of these carried the specimen of glass to be studied, the other entirely screened the slit from the radiation of the glower. To get a measure of the energy transmitted by a piece of glass for a particular wavelength, the deflection of the galvanometer when no absorbing medium was interposed between the Nernst glower and the spectrometer slit was divided into the corresponding deflection of the galvanometer when the radiation passed through the glass plate

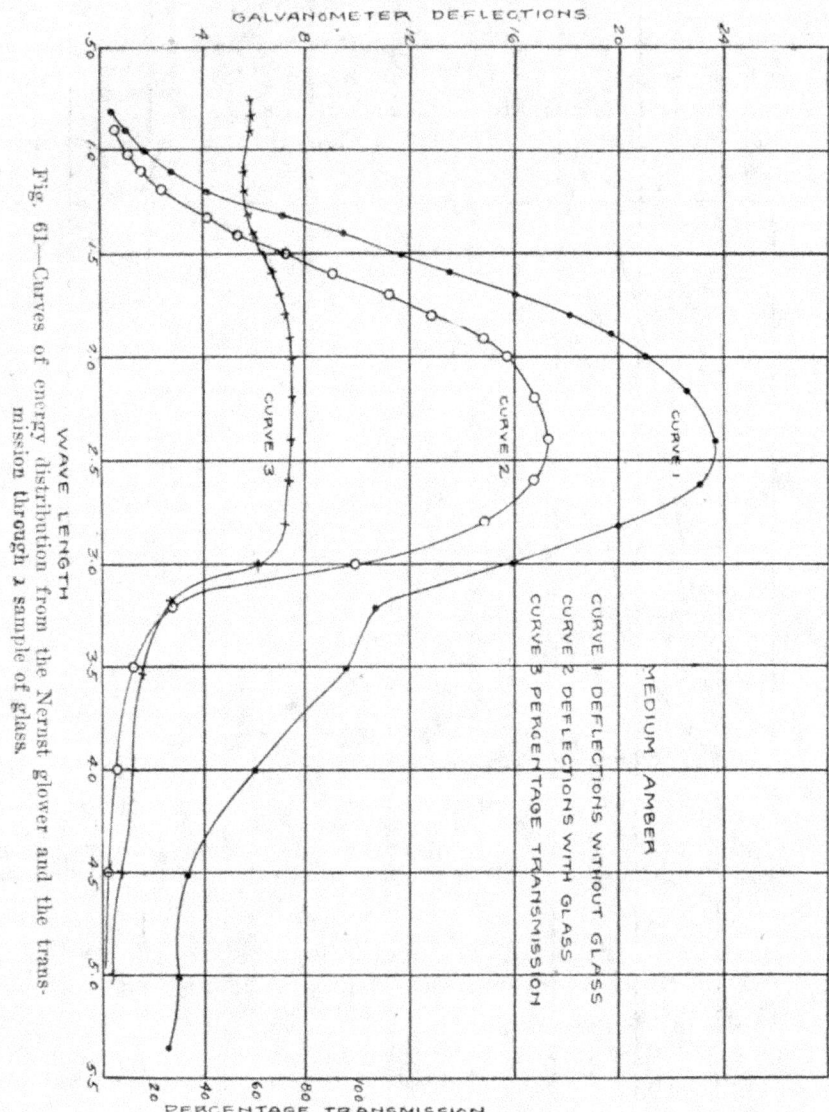

Fig. 61—Curves of energy distribution from the Nernst glower and the transmission through a sample of glass.

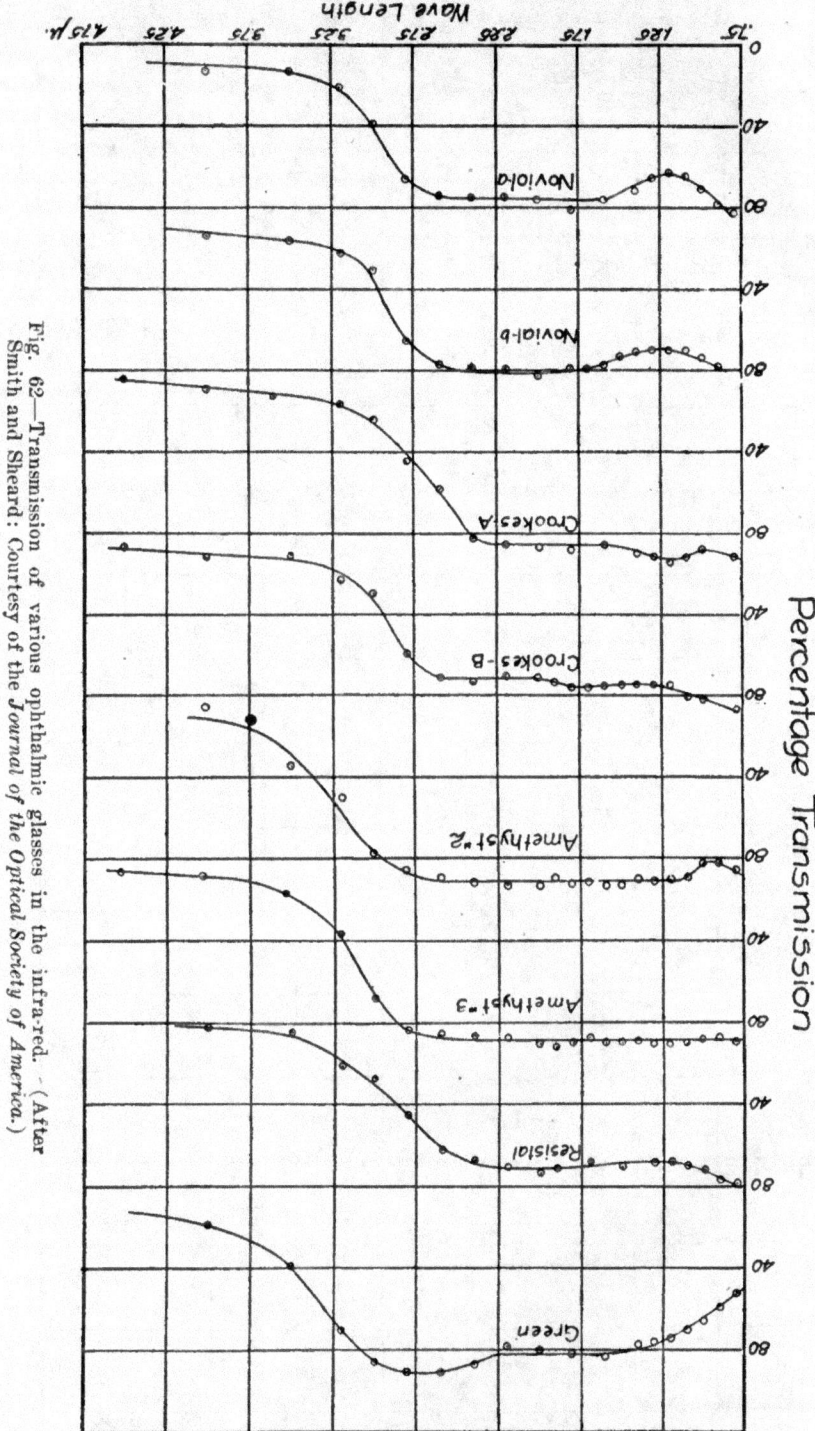

Fig. 62—Transmission of various ophthalmic glasses in the infra-red. (After Smith and Sheard: Courtesy of the *Journal of the Optical Society of America.*)

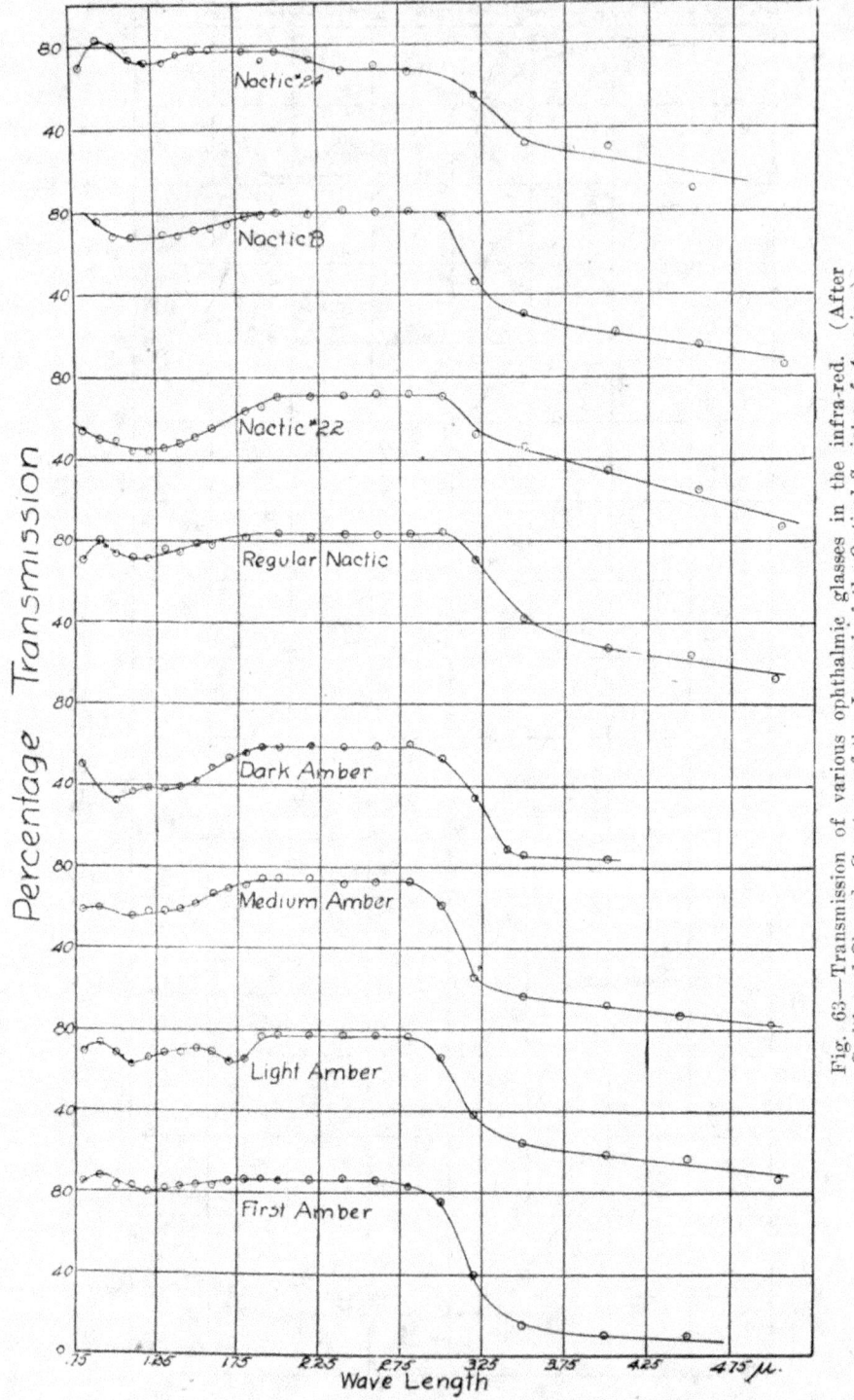

Fig. 63.—Transmission of various ophthalmic glasses in the infra-red. (After Smith and Sheard: Courtesy of the *Journal of the Optical Society of America*.)

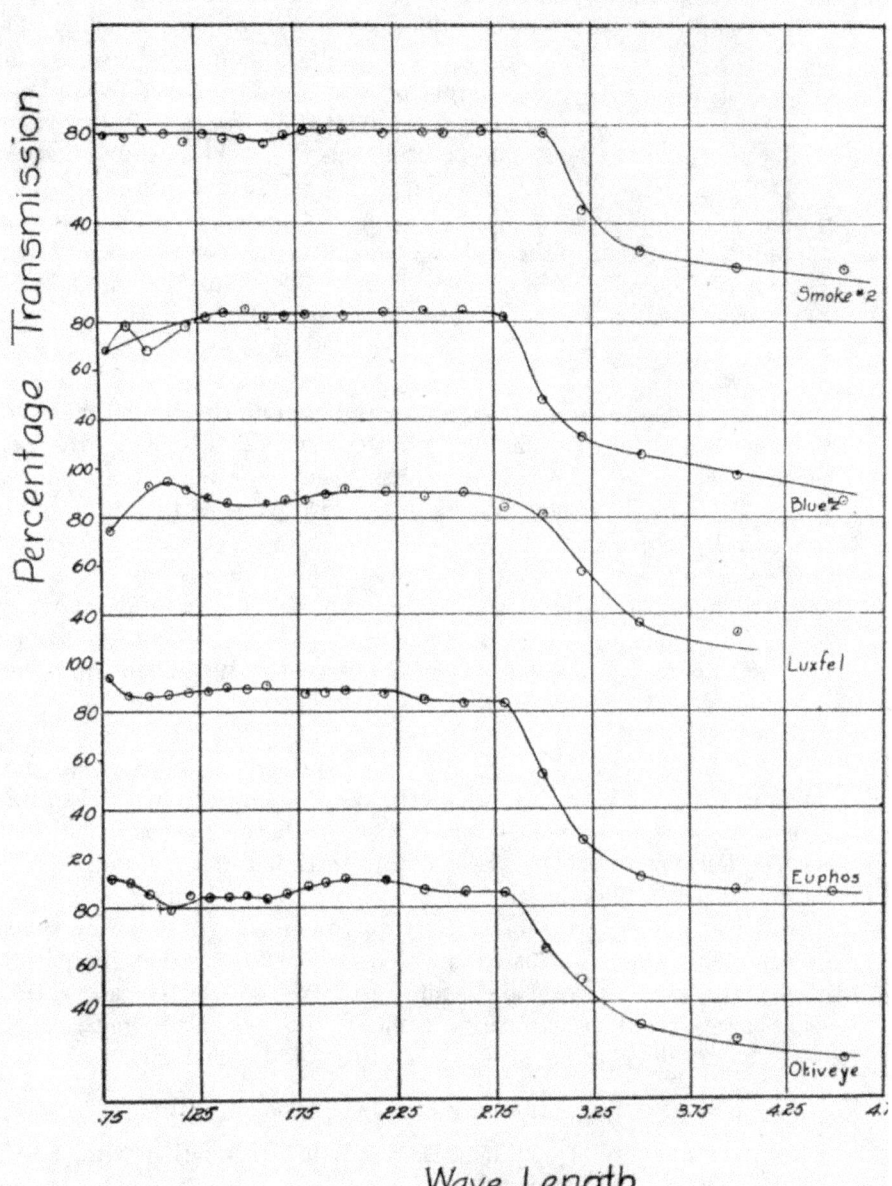

Fig. 64—Transmission of various ophthalmic glasses in the infra-red. (After Smith and Sheard: Courtesy of the *Journal of the Optical Society of America*.)

before it reached the spectrometer slit. Figure 61 gives a sample set of data obtained; the upper curve represents the energy distribution from the Nernst glower as measured by the galvanometer deflections and the lower curve, in a similar manner, measures the energy transmitted by the sample of glass. Figures 62, 63 and 64 contain a whole series of curves so related in general that various shades of the same colored glass are in sequence. The so called nactics —a trade name—and ambers occupy the whole of Figure 63. The rather noticeable selective absorption in the region of 0.75 to 1.75 μ (broadly) exhibited by the yellow glasses (noviol, nactic and ambers) is evidently characteristic of such glasses. The heat transmission of the German Euphos glass is greater than for any of the common ambers as tested except possibly that marked "First Amber" which it closely resembles.

In Figures 62-64 the wavelengths are plotted on the horizontal axis; the percentages of transmission on the vertical axis. In order to economize space the vertical axis has been so lettered that the 80 per cent. of one curve coincides with the zero point of the curve lying immediately above it.

It seems useless to attempt to present the transmission properties of all the numerous glasses which are obtainable under different trade names but which have a characteristic color. The color of the same kind of glass may differ somewhat for different melts and for different parts of the same melt. This may have a marked effect upon the visible spectrum but does not in general affect the coloring matter. Certain colors of glasses are difficult to match. There is, in our opinion, too great a variety of colored lenses and a strict standardization of these is to be hoped for. At any rate, as Verhoeff and Bell write, "Perhaps the chief benefit of the agitation that has taken place within the last decade on the possible * * * dangers of the ultra-violet has been the bringing into prominence of the new types of protective glasses. These, intended primarily for the elimination of the ultra-violet rays, have tended to types of selective absorption which give advantageous results in modifying the visible light, which is really the chief object of concern of the ophthalmologist."

CHAPTER IV.   TRANSMISSION OF THE OCULAR MEDIA.

In 1908 Parsons of London, England, in conjunction with E. C. Baly, F. R. S., an authority on spectroscopy, and E. F. Henderson carried out some investigations on the absorption spectra of the cornea, lens and vitreous of the rabbit's eye. Later Parsons and Martin made a more extensive study, the results of which appeared in the *Journal of the American Medical Association* (Vol. LX, page 2027, 1910). These researches were antedated in some particulars by those of Birch-Hirschfeld, Hallauer and Schanz and Stockhausen. The accompanying diagrams (Figures 65-69) are from originals taken at the Imperial College of Science and Technology in the laboratories of Sir William Abney and Professor Fowler, F. R. S.

The Ultra-violet and the Visible.

In the research under discussion experiments were made to determine the precise limits within which the short wavelengths of light

Fig. 65—The transmission of the cornea. (After Parsons.)

Fig. 66—Spectrograms showing the transmission of the crystalline lens.
(After Parsons.)

were absorbed by the refractive media of the eye and the effects on these limits of keeping the media some hours after the death of the animal. The media were mounted in cells with parallel sides In the case of the cornea and vitreous the cell was placed close to the slit of the spectroscope. The lens was dealt with in two ways:— (1) Suspended in normal saline and placed at a distance from the slit greater than its focal length, so that a blurred image of the source of light was thrown on the slit. In this way horizontal lines were avoided on the resulting photographs and the possibility of stray light entering between the cell and the slit was prevented. (2) A thin layer of lens substance was squeezed out flat between the parallel sides of the cell; this was done to eliminate any possible apparent absorption due to the shape of the lens.

All of the media were found to be uniformly permeable to rays between the wavelengths 6600-3900 t. m. For the ultra-violet rays the iron arc was the source and quartz was used throughout. Plates containing no dyes and giving no absorption bands were used. The results obtained by Parsons agree closely with those obtained by Schanz and Stockhausen and Birch-Hirschfeld. The shortest interval between the death of the animal and the taking of observations was three minutes. Observations were also made on the vitreous one hour, lens five hours and cornea several hours after the death of the animal.

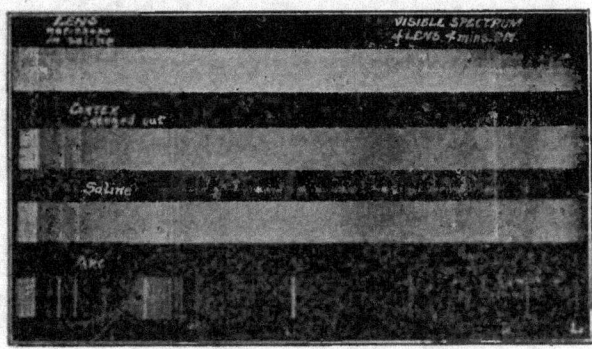

Fig. 67—Photograph showing the spectra of the crystalline lens in normal saline and the cortex. (After Parsons.)

The results obtained were identical with those from fresh specimens. The conclusions to be drawn from Figures 65-69 are:—

"*Cornea.* The cornea was found to offer no resistance to rays of wavelength longer than 2950 t. m., but all those beyond this limit were completely cut off.

Lens. (a) Suspended in normal saline. Rays of wavelengths less than 3500 t. m. are absorbed completely. The line is not a sharp one, the absorption commencing at about 4000 t. m. (b) Squeezed to different thicknesses. The absorption varies *pari passu* with the thickness of the layer of lens substance.

Vitreous. The vitreous in a layer $\frac{3}{16}$ inch thick shows a broad absorption band extending from 2800 to 2500 t. m., with a maximum at 2700 t. m. The margins of the band are ill-defined."

It would be unsafe, however, to apply those results obtained with rabbits' eyes directly to the human eye without further investigations. The results of Schanz and Stockhausen (various papers in *Klin.*

Monatsbl. f. Augenh.) on the transmission of the cornea and the vitreous of a calf's eye are confirmatory. Birch-Hirschfeld also investigated the transmission properties of the media of the eyes of calves, pigs and oxen. He discovered that there was little difference in the absorption of the ultra-violet by these corneæ, giving the limit as 3060 t. m., somewhat higher than for the rabbit and considerably more than that of ordinary glass. Birch-Hirschfeld found the limit of absorption to be 3000 t. m. for a layer of vitreous 1 cm. thick. Greater differences which cannot be overlooked were found with various lenses. The limits of transmission of the rabbits' lenses varied between 3300 t. m. and 3900 t. m. For the pig's lens the average limit was 3300 t. m. with variations of about 150 t. m.; for the calf's

Fig. 68—Spectrograms showing the transmission of the ultraviolet by the vitreous.
(After Parsons.)

lens at 3280 t. m. with variations of 120 t. m. and for the ox lens from 2700 to 4000 t. m. Schanz and Stockhausen examined the cornea and lens of a child who had glioma. The cornea absorbed up to about 3000 t. m. and the lens about the same. In a certain case of injury the corneal absorption was about the same but that of the lens was much greater, i. e., up to about 3500 t. m. Hallauer (*Klin. Monatsbl. f. Augenh.* 1909) found that the corneal and vitreous absorptions in the eye of a man extended to 2950 t. m. He examined the lenses from a considerable number of individuals of different ages and reached some valuable conclusions. He found the limits of absorption, on the whole, dependent upon age with some individual variations due to thickness, color and consistency. General conditions of disease also introduce a disturbing factor which must be taken into account in Schanz and Stockhausen's case of glioma. In babies and young children the absorption extends to about 4000 t. m. With this absorption, however, is combined an inability to absorb rays from 3000 to 3100 t. m. This transparent band is said to persist up to about the twentieth year and may be more extensive in certain debilitated conditions. Rather peculiarly, with the loss of this band after twenty, it is claimed that the limit of absorption drops from 4000 t. m. to about 3770 t. m. With advancing age, however, the crystalline lens becomes more and more 'yellow' as a general rule and therefore its absorptive powers reach down into the violet, extending even up to 4200 t. m. Extreme debility from disease diminishes absorption to a minimum of 3750 t. m.

All of these investigations show that the lens has a powerful capacity for absorbing ultra-violet light in the region roughly comprised between 3000 to 3800 t. m. The fact is very easily and strikingly demonstrated by the strong fluorescence which occurs when these rays strike it. Schanz and Stockhausen attribute this fluorescence to

Fig. 69—Transmission of the ultraviolet by the vitreous. (After Parsons.)

the rays between 4000 t. m. and 3500 t. m. In the paper *(Illuminating Engineer, 1910)* on Glare—Its Causes and Effects, by Stockhausen, we find this statement:—"Now the ultra-violet rays between 3750 t. m. and 3200 t. m. are strongly absorbed by the eye-lens and those between 4000 t. m. and 3750 t. m. are for the most part altered into fluorescence light in the lens. Violet rays also, as Schanz and Stockhausen have shown, generally contribute to some extent to this change. Now, in general, it is only those rays which are absorbed by any substance which exert a chemical action upon it and we are, therefore, justified in supposing that it is the ultra-violet rays which are absorbed by the lens that produce the effect referred to above. In addition, the conversion of all ultra-violet rays and a portion of violet light into visible light by fluorescence indicates a transformation of energy and in the course of years may produce the injury to the eye known as cataract." But, as pointed out by Helmholtz, a fluorescent body always strongly absorbs those rays which induce the fluorescence. Hence the chief rôle must be allotted to rays between 3500 t. m. and 3000 t. m., for those from 4000 t. m. to 3500 t. m. are absorbed to some extent by the lens. Also, as pointed out by Helmholtz and Stokes, the light causing a fluorescence is of a shorter wavelength than that of the emitted fluorescence. The investigation of this fluorescence of lenses is not unattended with complicating features, especially those due to fluorescence of the observer's own lenses.

The Infra-red.

The general absorption of the eye media has been studied by Aschkinass *(Ann. der Phys. und Chem.,* Vol. 55, 1895) in connection with his determination of the absorption spectrum of fluid water. He found that the transmission of the media of the eye for radiant energy in general was closely similar to that of water in a layer of equal thickness. The large proportion of water in these media would, of course, suggest a similarity and Aschkinass found the characteristic absorption bands of water in the experiments on the eyes of cattle and some control experiments on the human eye. The only notable discrepancy was in finding a considerably higher absorption in the cornea than would be warranted by its water equivalent. This Aschkinass ascribes chiefly to a film forming very rapidly over the surface of the dead cornea.

Hartridge and Hill, working in the Physiological Laboratory of Cambridge, England, have carried out some important investigations upon the transmission of infra-red rays by the media of the eye. This work is published in the *Proc. of the Royal Society of London*, Series B, Vol. 89, 1917. These investigators used a constant deviation Hilger spectrometer: in place of the eye-piece in the telescope there was inserted an adjustable vertical slit behind which was mounted a delicate thermopile of ten bismuth-silver elements. This thermopile was connected to a Paschen or Broca galvanometer and the energy falling upon the thermopile was measured by its deflection. The whole telescope was protected from radiant and convected heat by a silvered vacuum flask, the mouth of which was plugged with cotton. The light source was a single vertical Nernst filament. The spectral examination of the aqueous and vitreous offered no great difficulty mechanically, since they could be held in parallel-faced glass or quartz containers. With the lens and cornea this is not the case. Two methods are available: first, to dry the lenses superficially and then to squeeze them into a small trough. This method is not highly successful since the differences in refractive indices of various zones of the eye lens cause a series of confused images of the light source. A second and better method is the immersion of the uninjured lens in some fluid of suitable refractive index that will neutralize the convergence exerted by the lens on a parallel beam of light passing through it. Hartridge and Hill found that carbon tetrachloride was most suitable for this purpose; it has no absorption bands over the region to be investigated, it does not precipitate the proteids of the lens and has marked antiseptic properties. An extensive series of experiments proved that lens preparations made in this way gave the absorption bands corresponding to those of water. The absorption curve of water in comparison with the lens of the eye is shown in Figure 70. It will be apparent to the reader that a superposition of two curves showing the amounts of energy of different wave-lengths transmitted could not occur unless "equivalent" thicknesses of water and media were taken. The following table gives such data:

TABLE III.

Structure	Thickness.	Index.	Water (per cent.).	Equivalent thickness of water.
Cornea	1.15 mm.	1.377	90	1.04 mm.
Aqueous	2.5	1.355	99	2.38
Lens center	....	.....	84	....
Lens cortex	4.05	1.39	92	3.35
Vitreous	15.00	1.340	96	14.4

Such information is of great value since it permits the substitution of the equivalent thickness of water in experimental work, thus removing the tedium and uncertainty in results due to a time factor necessarily involved in dealing with anatomical media.

The table as given by Luckiesh (*Electrical World*, Oct., 1913) differs somewhat from the figures as given by Hartridge and Hill. Luckiesh's data are as follows:

Media.	Equivalent cms. of water.
Cornea	0.06
Aqueous humor	0.34
Crystalline lens	0.42
Vitreous humor	1.46
Total eye	2.28

The very important question arises: In what amounts do the infrared radiations of different wave-length gain access to the deeper structures of the eye? In other words, What is the energy density in the eye media? The answer to this question has been undertaken by Luckiesh (*Elec. World*, 1913) and by Hartridge and Hill (*Proc. Roy.*

Soc. of London, 1917). The intensity of radiation after traversing any depth, *d*, can be computed from the following equation:

$$I' = Ie^{-ad}$$

where *I* and *I'* are the original and final intensities respectively, *e* is

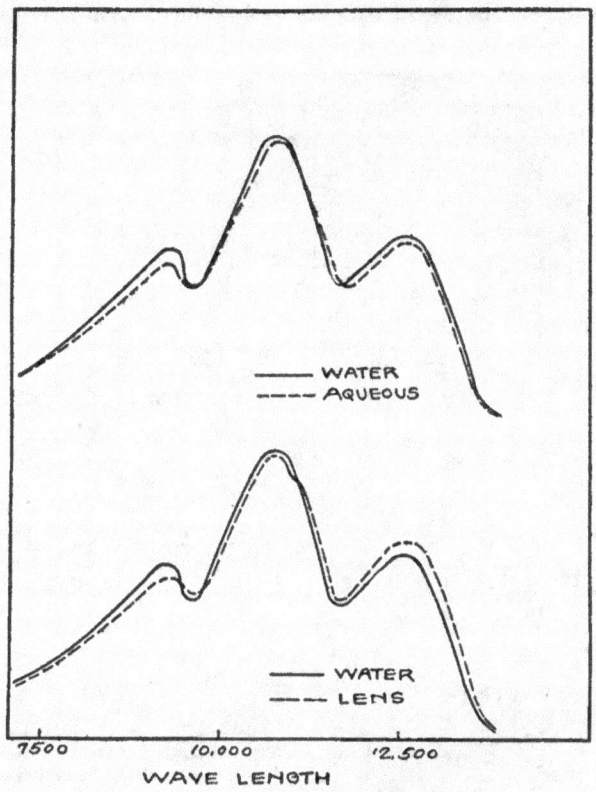

Fig. 70—A comparison of the absorption curves of water and the crystalline lens and the aqueous humour in the infra-red region. (After Hartridge and Hill.)

the base of the Naperian logarithms and *a* is the extinction coefficient. This can be further simplified for purposes of calculation thus:

$$I' / I = T = e^{-ad}$$

where *T* is the transmission coefficient. If *I* be taken as unity, then the value of *I'* is equal to that of the transmission coefficient. Aschkinass gives in his paper a table of extinction coefficients for pure water from $0.45\,\mu$ to $8.49\,\mu$. Hence it is possible to compute the transmissions of the various eye media within this range. Aschkinass did this for the whole eye: the transmissions of various layers of water corresponding to the eye media according to Luckiesh (*Elect. World*, 1913) are given in the curves of Figure 71. The first curve, that of the equivalent cornea, indicates the percentage of heat energy transmitted by the cornea of that incident upon the cornea; the second curve shows the percentage of heat energy reaching the anterior sur-

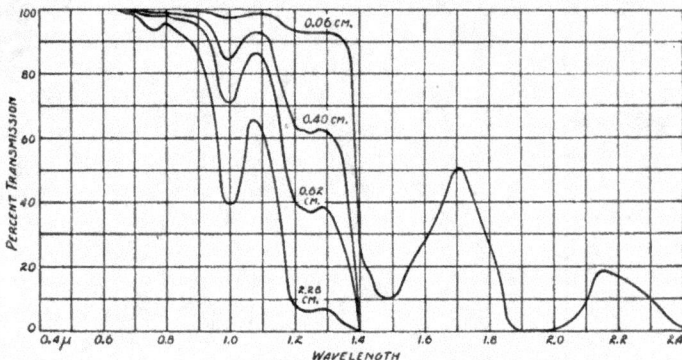

Fig. 71—Transmission of various layers of water corresponding to the eye media.
(Courtesy of M. Luckiesh.)

face of the lens of that incident on the cornea, and so forth. Table IV gives the set of experimental data obtained by Hartridge and Hill (*Proc. Roy. Soc.* 1917):

TABLE IV.

Wavelength.	Equivalent cornea.	Equivalent cornea and aqueous.	Equivalent cornea, aqueous and lens.	Equivalent eye.
7000	97.5	95	95	94.3
7500	97.5	95	94.6	91.3
8000	97.5	94.5	93.6	89.6
8500	97.5	94.2	93	89
9000	97.2	93.6	91.9	86.1
9500	94.4	85.4	76.2	48
9750	93.6	83.1	72.5	41.2
10000	94.5	85.8	77.2	50.3
10500	96.6	93	89	77.6
11000	95.9	90	85.1	67.7
11500	89.4	71.5	53.2	15.9
12000	86.4	63.7	42.2	7.9
12500	87	65.7	44.9	9.5
12750	87.3	65.6	44.8	10.6
13000	85.4	61	37.7	6.55

Wavelength.	Equivalent cornea.	Equivalent cornea and aqueous.	Equivalent cornea, aqueous and lens.	Equivalent eye.
13500	75	36.4	13.4	0.24
14000	23.5	0.7	....	
14500	5.5	....	....	
15000	12.9	1.2	....	
15500	28	1.38	....	
16000	48.2	8.7	00.73	
16500	53.3	12.2	1.44	
17000	51.4	10	0.95	
17500	43.5	5.6	0.3	
18000	20.3	0.4		
18500	4.9			
19000	2.0			
20000	4.4			
21000	7.6			
22000	5			

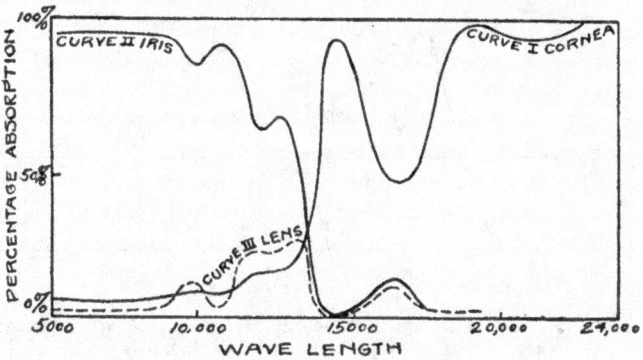

Fig. 72—Curves showing the percentages of infra-red radiation absorbed by the media specified of the amount of energy incident upon the medium named. (After Hartridge and Hill.)

These tables and curves show that there is practically no transmission of energy of wave-length greater than 23000 Angstroms. Paschen (*Wied. Annalen,* Vol. 52, 1894) showed that a layer of water 0.03 mm. thick transmitted at no wave-length more than twenty per cent. of the incident energy; a layer 2 mm. thick would, therefore, be totally opaque for wave-lengths greater than 23000 t. m. Furthermore, an inspection of Figure 71 shows that heat radiations of wave-lengths from 7000 to 9500 t. m. roughly pass into the eye almost unchecked and that a great deal of it reaches the retina. Figure 72 shows: Curve I, percentage of heat energy absorbed by the cornea of that incident upon it; Curve II, percentage of heat energy absorbed by the iris of that incident on the cornea and Curve III gives the percentage of heat energy absorbed by the lens of that incident on the cornea. The curves of this diagram are all representative of *absorption;* those in

Figure 71 give *transmission*. It will be noted that the absorption of the iris for wave-lengths ranging from 5000 to 10000 t. m. approximates 95 per cent. Hartridge and Hill (*l. c.*) say that the iris of the ox totally obstructs heat radiation of every wave-length which falls upon it. The lens, on the other hand, absorbs of the radiation which falls upon it by way of the aperture of the iris only about twelve per cent. Roughly stated, it can be said that four times the amount of energy is absorbed per unit area of the iris as is absorbed by the lens.

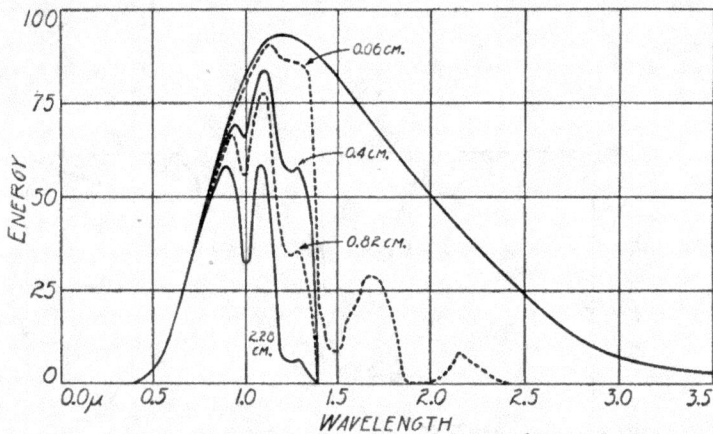

Fig. 73—Transmission of radiant energy from a 1.25 watt-per-candle tungsten lamp through various layers of water. (Courtesy of M. Luckiesh.)

Another point of interest is to apply these transmission curves to the curves representing the spectral energy distribution of black bodies at various temperatures and also to those of various illuminants. Figure 73 gives the transmission of radiant energy from a 1.25 watt-per-candle-tungsten lamp through various layers of water. (Luckiesh, *Elect. World*, 1913.) The numbers on the curves represent the thickness of water. For example, the percentage of total energy radiated from the carbon lamp and which is transmitted by the cornea is found by obtaining the ratio of the area under this curve (0.06 cm.) to the total area under the radiation curve. The difference between this and unity gives the absorption of the cornea. These percentages are found in Table V and plotted in Figures 74-76. Figure 74 gives the percentages of total black-body energy absorbed by the various eye media. It will be seen that for the cornea these percentages rapidly decrease with increase of temperature of the source, but much less rapidly for the aqueous, while the percentages of absorbed energy are at a maximum for the lens and vitreous humor at about 3500° K. Most of the energy is absorbed in the outer portion of the eye.

TABLE V.

Percentage of energy absorption.

Percentage of total energy absorbed in

Source	Water of depth				Cornea	Aqueous humor	Lens	Vitreous humor
	0.06 cm.	0.04 cm.	0.82 cm.	2.28 cm.				
Black body 2000°K.	68.8	80.6	83.8	89.7	68.8	11.8	3.2	5.9
Black body 2500°K.	51.7	63.6	68.3	76.7	51.7	11.6	5.0	8.4
Black body 3000°K.	38.5	49.8	55.7	65.1	38.5	11.3	5.9	9.4
Black body 4000°K.	22.8	31.7	37.2	45.9	22.8	8.9	5.5	8.7
Black body 5000°K.	13.0	19.6	23.4	30.4	13.0	6.7	3.8	7.0
4 w.p.c. carbon.....	64.1	77.3	81.0	87.9	64.1	13.2	3.7	6.9
1.25 w.p.c. tungsten.	50.4	64.5	70.5	80.0	50.4	14.1	6.0	9.5

Percentages of energy absorbed have only been considered. The data can be reduced to that of finding the actual watts absorbed per lumen. In Figure 76 are plotted the values of watts per lumen for the black bodies at various temperatures. Multiplying these values

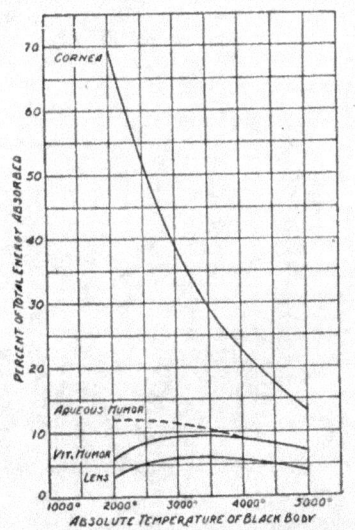

Fig. 74—Percentage of total radiant energy absorbed in various eye media.
(Courtesy of M. Luckiesh.)

by the corresponding values for the curves of Figure 74, the actual watts absorbed per lumen are obtainable. Figure 75 carries these results. Curve *a* represents the absorption for the total eye; curve *b* that of the cornea, and so on. These curves give the actual power absorbed in the eye media per lumen of light flux in the entering beam. All of the data show that the outer layer of the cornea absorbs a large portion of the energy which is not active in producing the sensation of

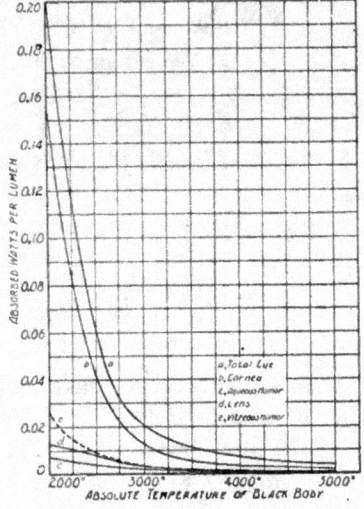

Fig. 75—Watts absorbed in the eye media per lumen in usual percentage of light. (Courtesy of M. Luckiesh.)

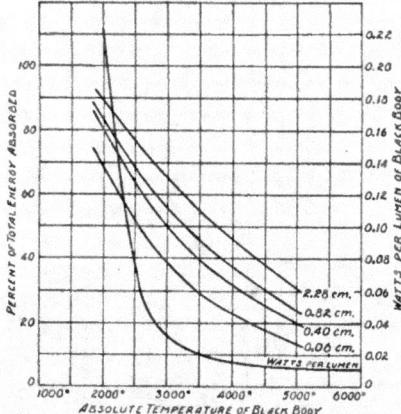

Fig. 76—Percentage of total radiated energy absorbed in the various layers of water. (Courtesy of M. Luckiesh.)

light. Also, as is to be expected, the absorbed energy per lumen of light flux incident upon the retina rapidly decreases with an increase of temperature of the source. It will be noted that about thirty times as much energy is absorbed in the total eye per lumen of tungsten light as per lumen of light from a black body at 5000° C. As Luckiesh says: "This same ratio would hold approximately for sunlight if it were not for the moisture in the atmosphere which absorbs much of the infra-red rays before they reach the eye. This is perhaps fortunate considering the enormously greater intensities of illumination encountered in daylight." For instance, according to F. E. Fowle (*Astrophysical Jour.* 1913) the amount of percipitable water existing in the form of atmospheric water vapor averages about 0.7 cm.

The marked difference between the action of water and the eye toward the infra-red on the one hand and the ultra-violet on the other hand is noteworthy. The eye media transmit the visible and infra-red rays in the same manner as water. This is not true for ultra-violet radiation. Water is transparent to short-wave radiation far into the

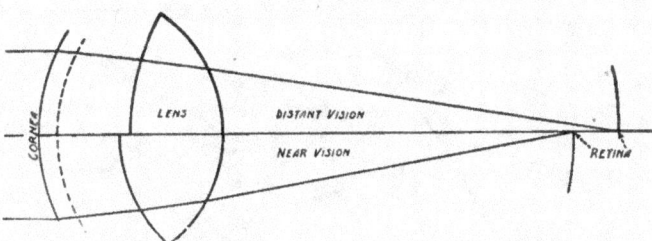

Fig. 77—Path of light in the eye. Small object. (Courtesy of M. Luckiesh.)

ultra-violet. In fact, no noticeable absorption has been found for any of the ultra-violet radiation from the mercury arc in a quartz tube.

The question of energy density in the eye media using sources subtending large and small solid angles has been discussed in a paper by Luckiesh (*Elect. World*, Sept. 1915). Figure 77 shows the path of light in the eye when a small object is looked at, while Figure 78 gives the path of light in the eye for an extended object. The useful beam of radiation included within a solid angle of 120° at the eye is shown by the full lines in Figure 78 when the eye is accommodated for reasonably near vision. If the object that is being viewed be illuminated with the same density of radiation of the same spectral character as that used for the small object tests at distance, it is obvious that the brightness of the retinal image will be the same and a much greater amount of energy will pass through the pupillary aperture. The energy density would thus be a million or more times as great as in the case of the more extended source. This is shown diagrammatically in Figure 79 for equal energy densities at the retina—that is, for equal brightnesses of the retinal images. Curve D represents the condition for the extended source and curve E for the small source.

In this paper Luckiesh says by way of summary: "It is shown that when viewing luminous objects of small area (subtending a small

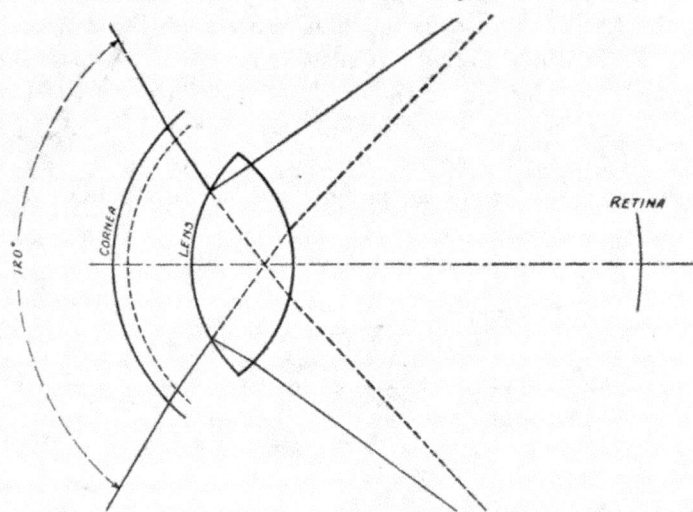

Fig. 78—Path of light in the eye. Extended object. (Courtesy of M. Luckiesh.)

solid angle) there is no serious concentration of energy in the eye media until the retina is approached. However, when viewing extended objects (large solid angle) there is a relatively much greater energy density in the lens and anterior parts of the eye than in the

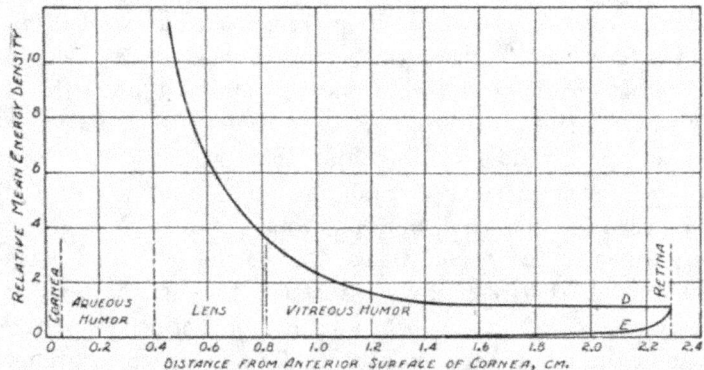

Fig. 79—Energy density in the useful beams of light from sources subtending large (D) and small (E) solid angles. (Courtesy of M. Luckiesh.)

posterior portions. When the retinal images are of the same brightness, there will be a much greater energy density in the lens when viewing an object subtending a large solid angle than when the object subtends a small angle if the spectral character of the illuminant and the intensity of the illumination are the same. This indicates that large sources of a relatively low visual brightness might be effective in forming cataract or causing eye fatigue if the "absorption of energy theory" is correct. In fact, if the deterioration of the lens is due to ultra-violet rays, the latter might be present in such small amounts as to appear harmless, but when it is recalled that the energy density in the lens is very high when viewing extended objects, such as the sky, pavements, large surfaces of molten glass, metal, etc., it appears to be possible that the ultra-violet rays might be present in sufficient amount to do damage. From this standpoint sunlight, owing to the greater intensities encountered, appears to be probably as effective in producing cataract and eye fatigue as ordinary artificial illuminants, even after allowing for the higher luminous efficiency of the former and the absorption of energy by the water vapor present in the atmosphere."

INDEX

www.ingramcontent.com/pod-product-compliance
Lightning Source LLC
Chambersburg PA
CBHW081211170526
45165CB00009B/2790